Common Science?

**Race, Gender, and Science**
Anne Fausto-Sterling, *General Editor*

*Feminism and Science* / Nancy Tuana, *Editor*

*The "Racial" Economy of Science: Toward a Democratic Future* / Sandra Harding, *Editor*

*The Less Noble Sex: Scientific, Religious, and Philosophical Conceptions of Woman's Nature* / Nancy Tuana

*Love, Power and Knowledge: Towards a Feminist Transformation of the Sciences* / Hilary Rose

*Women's Health—Missing from U.S. Medicine* / Sue V. Rosser

*Deviant Bodies: Critical Perspectives on Difference in Science and Popular Culture* / Jennifer Terry and Jacqueline Urla, *Editors*

*Im/partial Science: Gender Ideology in Molecular Biology* / Bonnie B. Spanier

*Reinventing Biology: Respect for Life and the Creation of Knowledge* / Lynda Birke and Ruth Hubbard, *Editors*

*Is Science Multicultural? Postcolonialisms, Feminisms, and Epistemologies* / Sandra Harding

# Common Science?

# Women, Science, and Knowledge

Jean Barr and Lynda Birke

Indiana University Press
Bloomington & Indianapolis

This book is a publication of

Indiana University Press
601 North Morton Street
Bloomington, Indiana 47404-3797 USA

www.indiana.edu/~iupress

Telephone orders 800-842-6796
Fax orders 812-855-7931
Orders by email iuporder@indiana.edu



The paper used in this publication meets the minimum requirements of American National Standard for Information Sciences—Permanence of Paper for Printed Library Materials, ANSI Z39.48-1984.

Manufactured in the United States of America

**Library of Congress Cataloging-in-Publication Data**

Barr, Jean, date
Common Science? : women, science, and knowledge / Jean Barr and Lynda Birke.
p. cm. —(Race, gender, and science)
Includes bibliographical references and index.
ISBN 0-253-33386-5 (alk. paper). —
ISBN 0-253-21181-6 (pbk. : alk. paper) —
1. Science—Study and teaching—Philosophy. 2. Science—Study and teaching—Social aspects. 3. Women in science. 4. Education, Humanistic. I. Birke, Lynda I. A. II. Title. III. Series.
Q181.B324 1998
500.82—dc21 97-47503

1 2 3 4 5 03 02 01 00 99 98

# Contents

Common Science?

# Acknowledgments

There are many people who have contributed to the development of our ideas over the years, and we cannot hope to thank everyone individually. Nonetheless, the research on which this book is based required the help of many people. Specifically, we are grateful to Eileen Aird of Hillcroft College, and to Gillian Hunt of Tile Hill College, and their respective staff and students. We also wish to thank those involved with running and studying through Warwick University's Open Studies Programme, as well as members of the Stoke on Trent Women's Health Group, Osaba Women's Centre and Cariba Women's Group in Coventry, the Henley College Primary Schools Programme, and the women from the Coventry Muslim community who participated in an ad hoc group to discuss women and science. Without the cooperation and help of each of these our study would not have been possible.

Individuals have also been important in discussing with us the ideas that we have developed in this book. So, we want to thank, in Britain: Flis Henwood, Sheelagh Doonan, Lorraine Blaxter, Liz Kelly, Kay Atkinson, Julia Berryman, Mary Brennan, Janet Alty, Jenny Whateley, Jane Conlan, and Julie Mills; also Lean Heng Chan and Rosalind (Ayisha) Whitworth. In the United States, Sandra Harding, Anne Fausto-Sterling, Ann Berlak, Evelyn Hammonds, Ruth Hubbard, Phyllis Robinson, Leila Ahmed, Sue Rosser, and Arlene Dallalfar. Thanks to all of you.

But most of all, we want to thank the various groups of women students with whom we worked, in adult education, women's studies, and teaching science to women in the community, who have helped us to think about what it means to teach science. And we particularly want to thank the women who graciously gave their time to discuss with us what they thought about science. Their generosity and articulacy told us much.

Common Science?

Common Science?

# Introduction

*"Science is everywhere, yet it has nothing to do with me"*

Most people must recognize that science* and technology permeate our lives, particularly in the West. New scientific discoveries are heralded in the media; computers have entered our offices and homes; our culture teaches us not to believe unless something is "scientifically proven."

Yet many people (and most women) stand outside science. The quote that heads this chapter summarizes a feeling expressed by many women—women who had little or no scientific training—in the course of our research and teaching courses for "lay" women in the community.

---

*Throughout this book, we use *science* to mean the natural sciences, as that is the most common use of the term.

What has happened to make these women feel outsiders to science and scientific knowledge? Partly, women's alienation stems from the way science is taught in school: few of the women we interviewed had much that was positive to say about science in their school days. Many women remembered only old-fashioned laboratory benches and bunsen burners—and poor facilities.

Partly, too, the distancing comes from recognition that science does not belong to women. Some women enter and excel in the scientific world, but for the vast majority of women (and many men) science is outside their experience. Training in science is the preserve of a minority, an elite: to study science or to become a scientist, it helps to be male, white, middle-class, heterosexual—and from an industrialized country. Science has been and remains a masculine domain.

At the same time, there is growing concern among western governments about the level of scientific understanding among the general public. In the United States, for example, indicators of public knowledge and understanding of science are published annually. In Britain, a wave of academic and establishment interest followed the publication, in 1985, of the Royal Society's report on this theme. One result of this in the United Kingdom has been growing government involvement; the British government has now set up a Parliamentary Office of Science and Technology and has been instrumental in setting up "Science Weeks." These are jamborees of activities that aim to publicize science and make it more fun and accessible to the public. Activities range from theatre and "hands-on fun with the kids" days, to more serious lectures and demonstrations.

While we welcome moves to communicate science to a wider public, we also have reservations. Science can undoubtedly be fun, but this type of approach can also produce a "gee-ain't-that-wonderful" response. This patronizes adults by ignoring the social and political contexts of science—contexts of which most adults are quite aware. And it is precisely these contexts which trouble many people. Visitors to Science Week may be left in awe, not only of the natural world but also of scientists. That is not necessarily the best, or the only, way to make science more accountable to a wider public, to make it more democratic.

The issue of democratizing science, and of women's exclusion from scientific knowledge, is fundamentally feminist. Modern science, as critics have often pointed out, is not sufficiently accountable to the public who pays for it. It is deeply Eurocentric as well as gendered, paying little heed to, or repudiating, other systems of knowledge (Harding, 1991).

Given the power of science in our world, that critique is important not only for feminist theories, but also for its contribution to wider struggles against the oppression of different peoples in the name of science. It is one of the "harbingers of hope" to which Hilary Rose refers (1994, p. 237), a way forward from the excesses of a culture obsessed with technology.

Alongside such critiques, concern about the lack of women and nonwhite people studying science is expressed by feminists and more "establishment" sources. Anxiety concerning the economy's need for a technically skilled labor force has contributed to the development of a wide range of initiatives designed to encourage women and minorities into scientific training. The journal *Science* regularly covers these themes. There are also initiatives that encourage women to take up scientific training and to stay with it. For example, Dartmouth College in New Hampshire has developed a mentoring program through which women science students are paired with a scientist who gives them support throughout their training. Similarly, the University of South Carolina carried out a model project to encourage the participation of women in science and mathematics (Rosser and Kelly, 1994). Contributors to Rosser's (1995) collection outline various schemes to encourage women in the physical sciences and engineering. These initiatives are doubly important; they help to encourage women into science and engineering, and they also help to identify the barriers that operate against women.

However, in the rhetoric surrounding this debate, it is often "women" who become the problem. The issue is perceived as how to get women or minorities into science: science itself is not usually questioned. From the perspective of feminist critiques, much is wrong with science and its practice. Science and how it is taught may well be part of the problem. Is lack of educational opportunity, then, the only reason why women don't enter science or why they don't always succeed when they do enter it? And is a job in the technically skilled labor force the achievement they really want out of it?

Our conviction, as feminists working in adult and higher education, is that the usual approach to women's absence from science fails many women. It fails partly because it does not take account of differences in learning style. The relatively rigid methods of science teaching might simply not match up to ways that girls or women want to learn (see Solomon, 1993). Why, then, should they choose to be trained formally in science, for a job? On the other hand, many people in our culture recognize the power of science and technology in the late twentieth century and want to understand more about what is happening. Parents, too, see their children learning things at school that were simply not understood in their own generation. The efforts to get more women into science may fail most women precisely because they are so firmly linked to the requirements of the labor force, thus denying other ways in which women might want to "make use of science."

Our perceptions of the gap between scientific knowledge and the lives of the majority of women are shaped by our experiences in the women's movement and by working as feminists in education. They are fuelled by our engagement with feminist critiques of science and our involvement with feminist groups that engage with science, such as health groups. Another contributory factor has been our work with women in the local community, developing initiatives based on science but not involving vocational training—science for women's interest rather than science for industry's sake.

Our research and teaching in the community are done in Britain, which has a long history of radical adult education. (The United Kingdom's tradition of elite higher education, rather than the mass access to higher education that characterizes the United States, is now changing.) Adult education grew out of the university extension movement and organizations such as the Worker's Education Association (which has links to the trade union movement). Its ideology has reflected those origins in that it was intended to extend educational opportunities to those who were unlikely to have access to elite institutions—in particular, working-class people. In practice, this meant working-class men, although some women did attend classes at the Mechanics' Institutes of the late nineteenth century.[1]

Adult education has been seen as offering the potential for a radical challenge to an elitist educational system. Based on recognition of people's experience, and on collective learning, the radical tradition emphasizes "really useful knowledge"—knowledge that serves people's needs rather than the abstract education typically offered in schools (Johnson, 1988). These aims accord with feminist praxis in the consciousness-raising groups of the 1970s and with women's health groups. They also accord with the practices of environmental groups opposing damaging developments, for example, Native American groups opposing the dumping of nuclear waste on their land, or environmental activists working in developing countries. What matters in these contexts is knowledge that will help the cause.

Consciousness-raising groups were important for the early development of women's studies in Britain. Science, by contrast, is not usually taught "from people's experience." It often bears little relationship to our everyday lives. This poses a problem for anyone interested in bringing science into educational work with community groups. How do we relate what we have learned as abstract knowledge to the experiences of the women with whom we work? "It's all right for you, miss," said a young working-class girl to one of us in a science classroom. Indeed. How can we make connections between the abstractions of science and that young woman's experiences?

The aim of our work has been to try to make sense of how nonacademic women perceive science. What understandings do they bring to thinking about science? What might they bring to any educational project? How are those understandings expressed in dialogue with us—white, middle-class feminists? Questions such as these were the starting point for our research with women students in adult and community education, and of our attempts to develop feminist pedagogy around science in the community. They also provide the starting point for writing this book. We want to explore some of these issues; to ask what women outside the academy might think about science, and about how their understanding might be shaped by their differing experiences.

There are always problems with doing research as a feminist. For example, in interviewing women, questions about power arise. We might try to have conversations with the women we interviewed, but they are always, at some level, interviews, and unlikely to be empowering or useful to the women themselves. On the other hand, such research with women may result in an understanding that can feed into other feminist work in the future.

We focus specifically on women. Undoubtedly, many of the processes of exclusion from scientific knowledge also affect men. Our concern, however, has been to work with women's groups, and that is our focus in research and writing. There are clearly issues about universalizing a "women's standpoint" (Harding, 1992; Rose, 1994); on the other hand, most women do share a status of being outside science to varying degrees.

We want to learn from the voices of the women to whom we spoke and from their different experiences—to begin to talk about science in terms of "speaking from women's lives." But we must recognize that we, as relatively privileged academics, are still interpreting the women's voices, however faithfully we try to represent the spirit of what each woman

says. And what we write here has to be set alongside the work of activist groups of women throughout the world who are challenging the outcomes of western science—environmental groups, workers' groups, and so on. Together, those strands constitute challenges to the hegemonic view of science as truth and as all-powerful. Those challenges are harbingers of hope for a better future.

# *"Science is too important to leave to the white coats"*

Science has brought us the benefits of antibiotics, but the horror of antibiotic-resistant strains of tuberculosis, too. No wonder the picture we encounter when we talk to women in our classes or in research interviews is that of a science that can go out of control, paying little heed to the needs of people or to their future. As one woman asked in an interview, "What kind of world are we leaving for our children?"

We can be fairly sure that it will be a world in which women have too little voice. Despite women's collective gains in the wake of feminism, there are many ways in which we continue to be silenced. For this reason, we base this book on the voices of women whose views and understand-

ings of science are not usually heard. We locate these conversations in the context of feminist writings about science and the nature of knowledge. The primary aim is to think about the implications of what the women have to say, both for us as feminist educators and for debates about the public understanding of science.

The world we live in is one in which the voices of the vast majority of people are not heard: "We are all part of science. We all take part in discoveries; we help one another find out why a tree is dying or what's wrong with us without going to an expert. And at home we figure out how to mend things. But our views aren't respected. Science is something everyone should know. It shouldn't be segregated. We'd be more confident if it wasn't left to the white coats" (Barbara).

For many of the black women, it is important that people help one another without the need to consult an expert (who is liable to be white). "But our views aren't respected," Barbara reminded us. "Race," too, determines whose voices will be heard in relation to science and technology.

Scientists sometimes make the claim that the public is "anti-science," or is "prey to propaganda." We doubt that public beliefs can be so simply summarized. Many people are anxious about the future of science, especially in relation to the nuclear industry or to developments in genetic engineering. The women we spoke to were acutely aware of the conflicts of interests that shape how science is done. But anxiety is not the same as antagonism. Concern for the future was repeatedly expressed alongside a realization that science could—if it was done for people instead of for profit and careers—provide lasting benefits and improvements in the quality of all our lives.

For women, access to scientific knowledge is problematic. That knowledge not only largely excludes women of all kinds; it also defines us—good reason why we should know more about it. Science is too powerful a kind of knowledge in late-twentieth-century western culture to be left to the boys. And it matters profoundly that the discourses of science contribute to particular definitions of gender—usually to stereotypical constructs of masculinity and femininity.

The women with whom we have worked regard science as a two-edged sword. They recognize that it can often lead to dangers and create environmental hazards. This is of particular concern because women are likely to have primary responsibility for children. On the other hand, they recognize science's benefits (though such benefits are largely experienced by people living in richer, more privileged parts of the world). They also recognize that the world is changing fast and that children are learning things in school that their mothers do not understand. For all these reasons, women want to find out more about science.

Two further aspects of women's relationship to science need to be addressed. First, nonacademically trained women expressed over and over again in conversation the profound elitism of science as a kind of knowledge. Science is something only "clever" people can learn. By implication, these women felt unable or unwilling to try to learn science, however important they may feel it to be. Some of the women also recognized that the kind of people who could learn it were likely to be men. This sense of being unable to do science is linked to their criticisms of science as something far removed from everyday life. If something seems to be irrelevant, why learn it?

Part of the perception of irrelevance may stem from seeing science's place in a dominant culture as its master narrative. Women stand outside this, particularly nonwhite women. Those black women interviewed who had immigrated to Britain relatively recently typically emphasized how "science" was not part of their original culture; indeed, science denigrated the understandings of the world brought from their own cultures.

A critical step in women's education is to find ways of teaching/learning science that do not perpetuate these feelings that it is an elite, and fundamentally difficult, way of knowing about the world, or that it bears no relationship to nonwestern cultures. This, of course, begs the question of whether or not it matters if women know much about science. We think it does because of its nature as an elite form of knowledge. That needs to be broken down; we can all take part in creating knowledge.

If this is to happen, scientists and science educators have to learn to listen to the voices of those who have been silenced. Different women bring different perspectives to bear on science. Science education has become elite because it has been based on the experience of a tiny minority: other people's perspectives and knowledges, other people's sciences and ways of discovering the world, have simply been ignored. And they continue to be ignored. Even within the rich countries, the accumulated wisdom of people who work alongside nature (farmers, for example) counts for nothing: only the "proof" offered by scientists can count.

There was no perception among the women to whom we spoke that science could possibly be enjoyable. On the contrary, for many women it was a litany of facts to be (passively) learned. This perception sits curiously alongside the awareness of the power of scientists. Of course,

science has contributed to many problems we face today, as well as to developments we might value. But learning it and doing it can be enjoyable, as some children find out. And if it is fun, it is less distant from our lives—which in turn might make it easier for people to challenge the "experts" in the wider political domain.

These are themes that we explore in this book. Many of the issues raised here apply to other fields of inquiry; we are particularly interested in science, and so that is our focus. Science is, moreover, a kind of knowledge that has high status and authority in our society. It is gendered and culturally specific. For these reasons, science is an important matter for feminists to engage.

## Feminist Critiques of Science

Feminist critiques of science have tended to have two broad themes. The first theme is concerned with who does science, recognizing that *science* (as we usually understand the term) has largely been done—and defined—by a tiny minority of people. Feminist critiques have shown how the involvement of other groups of people with science has been written out of history (see Harding, 1992; Keller, 1985). Today, women's names are not on the whole associated with the practice of science. Even among feminists it is sometimes hard to find anyone who can name more than one famous woman scientist (and almost invariably this is Marie Curie). Similarly, people from less industrialized countries are much less likely to have their names associated with science than are men in the West. So feminist literature has been concerned with the extent to which women and others are underrepresented in science and to look at why that has come about.

Science, feminists have emphasized, is very much part of the wider culture. Feminist critics have looked at the history and philosophy of science,[2] at particular aspects of biology, and at the intersections of sexism and racism in science.[3] Feminist literature on science, however, reflects the development of women's studies within the academy. As a result, the voices that we hear within that literature are largely those of academic feminists.

Yet within the broader women's movement there is a history of women engaging with issues that have to do with science. Women have actively resisted some reproductive technologies, for example, and have worked in the environmental movement in ways that engage with the content of science. In women's health groups, especially, women have been working not only to oppose mainstream orthodoxies, but also to construct their own knowledge—knowledge that in other contexts might well be called scientific. In this sense, the women's movement, like the environmental movement, has had an ambiguous relationship with science. It has been critical of, yet at the same time has made use of, scientific knowledge.

## Women's Exclusion

The example of women's health groups demonstrates that despite ambiguities, finding out how our bodies work and creating knowledge has been a radical act. It challenges prevailing medical assumptions that we are simply "patients" to whom things might be done. Nevertheless, outside the medical context women have in profound ways remained largely excluded from scientific knowledge. Working with women's groups and women's education raises questions about access to kinds of knowledge that are associated with power. Science is quintessentially a powerful form of knowledge within our society, and women are in practice disenfranchised from it. That matters, both for women's development as citizens within an increasingly technological society and because obtaining access to knowledge of any form can be empowering.

To be able to construct knowledge, of course, or to perceive that one can, is even more empowering. This point is emphasized by Mary Beth Belenky and her colleagues in their study of "women's ways of knowing" (1986). But if, as we suspected, most women feel completely alienated from science (a feeling inculcated by the methods of teaching science as a series of facts in the first place), they are unlikely to do more than passively accept scientific knowledge. In turn, passive acceptance of knowledge cannot confer much sense of ownership of knowledge, nor, indeed, of science as an authoritative voice. So, in our research, a first step was to find out how women see their knowledge, or lack of it: to find out how they locate themselves in relation to science and its practice.

In approaching our research, we already knew from our classroom experience that women often feel they have not, and cannot, play a part in the generation of such knowledge. We knew, too, that the knowledge that women do have (that might be considered scientific in other contexts) is often knowledge associated with women's role in the family—knowing about food additives and nutrition for instance. This raises many questions about how such knowledge is acquired/constructed.

Such observations are not, of course, unproblematic. There is clearly a danger that in recognizing and acknowledging these links we put ourselves in the position of implicitly supporting women's domestic role. Yet that role is (at least partly) what constitutes the everyday experience of many women. What concerned us more, however, was that any knowledge that women had, they themselves were likely to see as not being scientific—just because it was what they knew.

Women who are beginning to study science as adults say that one of their reasons for doing so is to keep up with their children at school. Others emphasize their awareness of the ways in which science and technology are developing so rapidly in the industrialized world; they want to know about what is going on around them. They want access to that knowledge, even if they know that they aren't going to be creating it in detail in the way that scientists do.

Science education in both Britain and the United States has developed largely as a form of abstract knowledge, removed from the everyday lives of people. Yet in both countries, in different ways, there has been a history of science education that included efforts to democratize it, to extend it to working people (though not without opposition: see Layton, 1973, on British science education, and Montgomery, 1990, on science education in the United States). Indeed, this was part of the agenda of the Mechanics' Institutes in Britain in the middle of the nineteenth century that were set up primarily to educate working-class men. There have been, too, many women popularizers of science.

The move to make science education more abstract was, ultimately, political, the establishment of an elite kind of knowledge (Layton et al., 1986). Science education became increasingly linked to class and gender interests. Science for what has been called "specific social purposes"—a more popular and accessible kind of science—began to disappear from the educational agenda by the late nineteenth century (Layton et al., 1986). In the United States, claims were made during that century that science education should be for all, for the "good of the common man" (Montgomery, 1994, p. 66). But here, too, more academic approaches took over during the early twentieth century, as the importance of scientific research for national development became clear. With the rise of academicism, science teaching became a matter of repeating "great" experiments. As Montgomery argues, "The 'science' popularized by this type of teaching, in an era of mass education and growing competitiveness, was thus an inevitable and heroic march upward to the present, enacted by lone investigators employing the tools of genius . . . the student commonly recapitulated this simplified and purified 'science,' which thus remained a coveted professional image for decades to come" (pp. 154–55). As we shall see, these images of science, scientists, and science education have remained salient cultural icons.

Contemporary debates about the "public understanding of science" (or scientific literacy) rest on a historical background of popularization of science and the extension to the public of developments in scientific knowledge. Yet this wider context is mainly absent in science teaching in the West today; instead, we have seen an emphasis on the need for scientists to popularize science. This comes from a scientific establishment concerned less with the benefits to the citizenry than for the need to defend its own interests. The concern here is not with empowering people but with making sure that they know what the scientific establishment wants to tell them. Dorothy Nelkin, in her book *Selling Science* (1987), an analysis of the ways in which science popularization (and particularly science journalism) operate, has made this point quite clearly.

For most women, wanting to become scientists is simply not part of the agenda. Yet at the same time (particularly in relationship to experiences of their body) they encounter scientific knowledge as objects of its inquiry. In her recent book *The Politics of Women's Biology*, feminist biologist Ruth Hubbard quotes the Brazilian educator Paulo Freire, who emphasizes what he calls "the indispensable unity between subjectivity and objectivity." This unity, Hubbard suggests, is what feminist methodology is all about. Science, by contrast, relies on the myth of complete objectivity. Hubbard points out that this is a form of context stripping, in which whatever it is that science studies is devoid of any location in the world. Such a denial of that indispensible unity between subjectivity and objectivity removes science completely from the everyday world and experience of people.

How can women, or others who are "outside" science, get beyond the stripping and discover the context? Feminist self-help groups focusing on health have provided one practical avenue for women to obtain information about how their bodies function. The generation and sharing of knowledge in these groups has enabled women to gain access to medical knowledge that has previously been denied them. On the other hand, how women gain the knowledge, and where it comes from, also raise issues for feminists. Information leaflets produced by medics are sometimes patronizing, and the very fact that they are produced for "the patient" by the

doctor reinforces relations of power. Moreover, the way in which information is written up may convey all kinds of assumptions, even in feminist texts. Susan Bell (1994) notes, for example, how the earlier versions of the Boston Women's Health Collective's *Our Bodies, Ourselves* unwittingly perpetuated scientific narratives that portrayed menstruation as deficiency. She recounts how later versions attempted to change the language to create more positive stories of how women's bodies work.

Most people have only limited access to science. Perhaps more importantly, many people believe that they can have only limited access or understanding. This belief is a result of years of exposure to an educational system that marginalizes people whose knowledges do not "fit" establishment norms.

In writing about how conventional education perpetuates existing patterns of social power, Paulo Freire discusses the need for a radical education that empowers people, distinguishing that from the "banking system" of education that prevails today. Perhaps in no area of inquiry is the banking system more predominant than in science. In order to study science at higher levels, people are expected to have "acquired" particular information, to have assimilated the requisite number of "facts." These in turn are assumed to be value neutral. Thus, for example, the deficit model of menstruation becomes a mere recitation of the facts of how the bodies of human females work. Yet, as a way of seeing women's bodies, it is imbued with socially constructed values. Freire recognized the way in which such education disempowers people. Women, in particular, may not acquire the requisite "facts" because the requirements themselves tend to assume previous experiences that are not theirs. What lurks behind the calls for greater public understanding is a banking shadow: if only "the public" knew more of the facts about science, we are assured, then they would understand and accept it better. Only then would the assets of science be properly protected.

This view of lay people as deficient in their knowledge of "the facts" disempowers those who are defined as lacking these facts. In writing about her own education, bell hooks (1994) has described the painful transition from an experiential education to the distancing and disempowering pedagogy of elite (white) institutions: "School changed utterly with racial integration. Gone was the messianic zeal to transform our minds and beings that had characterized teachers and their pedagogical practices in our all-black schools. Knowledge was suddenly about information only. It had no relation to how one lived, behaved" (p.3). Later, she adds that "I learned that far from being self-actualized, the university was seen more as a haven for those who are smart in book knowledge but might be otherwise unfit for social interaction" (p. 16). That is a theme understood by many of the women we interviewed, for whom their own knowledge—be that "common sense" or skills in "social interaction"—was something to be valued in contrast to the factual and alienating knowledge of science.

These women's responses emphasize how important it is to recognize and validate the kinds of knowledge that women do have. Official pronouncements about "public understanding of science" have largely failed to consider how and why people understand things. Many such pronouncements have perpetuated the image of a "public" sadly misinformed and ignorant of science and technology—creating what has been termed a "deficit" model of the public (Wynne, 1991). From a feminist perspective, that model is likely to be gendered (who is most likely to be seen as deficient in knowledge?) and to ignore the wider context of knowing. It also, significantly, ignores the way in which subjectivities are denied by formal educational practice—especially in teaching science.

This is one of the important themes of our research that focused on women's perceptions of science. Women we interviewed find this context-stripping of science absurd, unreal, and divorced from the reality of their own lives. Scientific knowledge might well be contrasted by some women to "common sense." Common sense, we were told, is what a woman might know already; that is, it is owned knowledge. This theme is echoed in other feminist work; Wendy Luttrell (1989), for example, notes how working-class women claim their own "common sense" as a superior kind of knowledge which is of more use than anything learned from books.

Scientific knowledge, however, is that which is too difficult to understand. "If I understand it, it can't be science" was a refrain in the research. The equation of science with extreme difficulty was best summed up by the interviewee who said that "getting through a day is like a science for most women."

## Promoting Women's Access to Science?

If people are outside science, how do they acquire scientific knowledge if they need to? Clearly, people do find out what they need, say, if they want to oppose a radioactive waste dump in their town. Less specifically, an obvious way in which adults might get access to scientific knowledge is through formal education. Partly because of a changing political climate, vocational courses for adults are proliferating in Britain. Industry's need for qualified scientists has woken people up to the existence of women; retraining courses in science have developed to encourage mature women to enter scientific careers.

Yet these openings do not necessarily meet women's needs. The women we interviewed were not interested in scientific training; rather, they sometimes felt that they were several steps away from any decision about their future careers. Many of them had only recently taken the huge step in their lives of returning to education—with all the terrors and anxieties that step might bring. The outcome of that education, and possible career plans, were light years away for them.

These women were all involved in adult education and/or working in community groups with educational interests. As such, they had an interest in seeking knowledge; they had at least taken that first step. Yet for many, it was enough—for now—to get back into education. Where that takes them in the future is not something they have mapped out. The shift toward "vocational" training is not something that is going to help—or empower—their lives.

Formal vocational courses will not suit women who simply "want to find out," to expand their horizons. So where do they turn? College courses are too geared to vocational outcomes. Science-based courses in liberal adult education are few, and when offered often fail precisely because so many people feel alienated from science. What, then, does someone do if she or he wants to seek science knowledge?

One of the more interesting developments has occurred in the Netherlands, in the form of "science shops." These centers are attached to universities and act as a liaison between the research expertise of the universities and various community groups. They have been in existence for some twenty years and appear to work reasonably well. However, a similar experiment in France seems to have failed, for reasons that are as yet unclear. One commentator has suggested that it may be because individuals have been able to use the science shops, and the shops have been subjected to the idiosyncrasies or whims of particular individuals (Stewart, 1988). In the Netherlands, community groups can approach the science shops in order to obtain information, or if the information is not available, they are put in contact with interested researchers and a joint research enterprise may be set up.

Two enterprises were set up in Britain in the late 1980s along the model of the Netherlands science shops. However, unlike the Netherlands, where women's groups do use the shops, at least one of the British equivalents has stated that women rarely approach it. We might speculate on the reasons for this in terms of women's exclusion from scientific knowledge. Nevertheless, the science shops do at least have the potential to become an interesting and radical experiment in democratic science education.

Something like a science shop also offers the potential for educators to encourage women (or others who are also disempowered by science) to find ways to study science for its own sake, or—more importantly—for their own sakes. Women could find things out for themselves, or work with someone who helps them to begin their own research. This would be much more what Sandra Harding (1991) has called "Developing Scientific Knowledge in Relationship to Starting from Women's Lives." Such an approach contrasts with the "getting women to become scientists" view (allowing them beyond the hallowed portals of elite science), which is not starting from women's lives.

## Starting from Women's Lives: Similarity and Difference

However much we recognize the importance of differences between women in "starting from women's lives," the existence of feminism presupposes some common ground. One theme we explore in interviews is the shared ground of women's exclusion from science. At the same time, the different experiences of the women will inevitably help to shape their differing perspectives and understandings, thus providing another theme. For example, we might expect to find differences relating to race or to social class. We also focus on the different perspectives that are shared within different groups of women, reflecting the milieux created within their community groups or adult education classes. These "group" differences are a critically important part of learning and have not been paid enough attention in thinking about pedagogy. They help to create what have been called epistemological communities—communities of learning and knowing—that become part of how we know about our world.

There are also differences between the social situation in Britain and in the United States that we might expect to find reflected in interviews conducted in the United Kingdom. One example is the position of black people, which has historically been rather different in the two countries. Many members of the black community in Britain arrived as part of the 1950s' wave of immigration encouraged by the British government to supply the labor force. Some of the older Afro-Caribbean women in our sample were born in the West Indies and have made a huge cultural transition during their lives. (In this sense, their experience may be similar to some of the Hispanic population who have come to the United States relatively recently.) Social class, too, has different meaning in the United Kingdom and the United States. British culture on the whole still places a great deal of emphasis on social class, to an extent that social class is seen as more rigid and less amenable to change than it is in American society. All these differences affect how women relate to knowledge.

## Identifying Tensions

A number of tensions, expressed during the interviews, reinforce the view that women are alienated from scientific knowledge. We will examine these in more detail in later chapters of the book; here we introduce them briefly. One example is the tension that we have already referred to, the distinction between common sense and science; that is, common sense refers to the knowledge that is owned, acknowledged, whereas science is that which women claim that they do not know.

Another tension is the distinction between using science and understanding the principles. Several women point out that although they may use science and technology in their everyday lives this is not the same as understanding the underlying principles, from which they feel very much removed. The third distinction that comes out of the interviews is that "science is all around us," but "it has nothing to do with me." Nearly all the women acknowledge that science is in everything—there are scientific explanations for all kinds of things, from the sun appearing to rise in the morning to why bread rises when it is baked. This acknowledges the power of science in our culture to explain the world around us. While recognizing that science is an important part of our lives, they also locate themselves outside it by consistently saying that science has "nothing to do with me."

Those kinds of dichotomies are of great importance to anyone concerned with the empowerment and education of women. They are of concern to anyone who is interested in the ways in which women are disenfranchised by the construction of scientific knowledge in our culture. And they are of great importance to feminist critiques and discussions of science.

Yet more optimistic strands do emerge from the research interviews. Certainly, the belief that science is all around us is one strand, even if it is contrasted with the notion that science has "nothing to do with me." If women believe that science is all around us then they recognize its power in our society. And that is a political recognition. Often, too, women would both set up the dichotomy and challenge it themselves, asking us if we had noted their contradiction. This suggests that they understand the political framework of science. It is important to stress here that such political awareness is a far cry from the image of women's "ignorance" of science that emerges from the "public understanding" rhetoric.

The women are also understandably skeptical of expert advice and of notions of proof and disproof. Although they sometimes locate themselves outside scientific knowledge, at other times many emphasized their own values as knowers, and that is important, too. As feminists, we must emphasize that women are not simply passive knowers; they may embrace ignorance of science knowingly, or they may resist the scientific account.

In the final chapter, we use those research findings and what they tell us about women's perceptions of science to draw out implications. What are the implications for education, particularly of adults? What are the implications for feminist critiques of science, or for feminist campaigns around scientific issues? Finally, if women do want more access to scientific knowledge, how can that be facilitated? Or perhaps we should phrase the question differently: how can women become involved in the creation of scientific knowledge, and how can we facilitate the validation of that knowledge within a wider culture?

## The Quest for the Holy Grail

Advocates of science use strong rhetoric. In *The Book of Man*, world-famous geneticist Walter Bodmer writes:

*This [is] the story of one of mankind's greatest odysseys. It is a quest that is leading to a new understanding of what it means to be a human being, and is now being carried out under the auspices of the Human Genome Project. . . . it is biology's answer to the Apollo Space Programme, yet it would have seemed audacious—if not downright preposterous—if it had been proposed only a decade ago. (Walter Bodmer and Robin McKie, 1994, Introduction to* The Book of Man*)*

Governments are understandably concerned about how the public perceives science. Scientific research costs money—sometimes a great deal—much of which comes from taxes; the Human Genome Project, for instance, is supported by billions of federal dollars. How the public perceives scientists matters, too, to many scientists who are concerned about their public image. Perhaps nowhere is this more acute than in relation to areas of controversy—genetic engineering is one example, the use of animals in experiments another. The above quotation illustrates the force of rhetoric in presenting debates to a wider public in the context of "popular" books: Walter Bodmer, a geneticist with a keen interest in promoting public understanding of science, prefers to portray new developments in mapping the human genome as though it is the search for the Holy Grail. Science, we must infer, is wonderful. Feminist critics, on the other hand, such as Ruth Hubbard, point out that this search calls into question "what it means to be a human being" and how that might be threatened by the new genetics.

How do nonscientists see these issues? What might they know about the relevant science? And how might they interpret conflicting messages and pictures of science? The context is one in which governments of industrialized countries express concern over public failures to understand much science; it is also one in which public perceptions can be "up for grabs" through powers of rhetorical persuasion. How scientists express themselves in public pronouncements is an important part of communicating science. It is what might be called the "body language" of science. The tone of a message can convey as much about science as the message itself.

In 1985, the Royal Society of London produced a report on the public understanding of science that set in motion a flurry of research activities and policy initiatives in Britain. Its main focus hit the headlines: The Public Knows Too Little Science. The statistics, gained from public surveys, make depressing reading at first glance—there seem to be few people who can claim to understand much about what is going on in science today. In the wake of that report, the Royal Society created a Committee for the Public Understanding of Science (COPUS). The British government produced its own report and established its Office of Science and Technology.

Among other things, the brief of this government office is not only to consider public understanding of science, but also to act as an information resource for politicians.

According to the Royal Society's report on public understanding, a variety of reasons for concern exists. These center partly on economic issues such as the need for a technologically developed society to have a technically trained workforce. They also focus on science as a part of culture—both educationally and aesthetically it is desirable that people know some science. Another focus is on science for citizenship—knowing something about science can facilitate our engagement with technical decision making.

To some extent, these are worthy aims. Yet each implies a different context of understanding and perhaps different public(s). Exactly who is "the public" that is invoked? Is it you and me? Is it the woman at the supermarket register or the farmer ploughing the field? Is it the people who run the town council? Does it include national politicians? And does every member of "the public" have the same need to understand science—what about working-class people? Nonwhite communities? None of these people are "scientists" in the way most of us might imagine (and that in itself begs the question of who, or what, is a scientist), yet each might have or need different understandings of (or about) science.

*Understanding* and *science* are also tricky terms. Official pronouncements about "public understanding of science," are usually concerned with ways in which "the public" is said to lack knowledge. But "understanding" might also mean an understanding of the role played by science and scientists in our society. Aside from the problems of definition, the aims noted above also fail to address questions of power and so fail to deal with issues of gender. And, of course, they fail to address science in its global context.

Our concern here is with the needs, lives, and desires of citizens, and particularly women, rather than with the economic "needs" of capitalist nations. Why should it matter if people fail to understand science? In this chapter, we explore some of the issues raised by research into "public understanding of science," particularly for feminists and for feminist pedagogy. We will also consider ways in which these debates link with women's understanding of science. While we would like women to have more access to scientific knowledge, we do not share the perspective implicit in government pronouncements, namely, that women (or "the public" generally) need to know more facts. The understanding that we both recognize in our conversations with women, and want to encourage, is not a passive understanding but a critical, active one. This is an understanding that does not take the facts of science for granted, but—as feminist critiques do—begins from a questioning stance.

## Surveying the Public

Official concern with the public understanding of science—or its lack—has been fuelled by various large-scale surveys into people's attitudes toward, and understanding of, scientific information. Among the questions asked in the United States surveys (published in annual *Science and Engineering Indicators*) were a set which, the researchers felt, tapped into people's understanding of what it means to do science or to study something scientifically (for discussion see Bauer and Schoon, 1993, and response by Miller, 1993). Perhaps predictably, few of the respondents achieved a high score on measures of scientific understanding.

A parallel survey in Britain in the 1980s produced similar results. When people were asked about interest in scientific discoveries, there was little difference between the British survey and similar surveys in the United States: approximately 40 percent claimed to be interested and 15 percent claimed no interest. There was, however, a striking contrast between the two countries in terms of self-professed interest in new medical discoveries. In the United States, 71 percent claimed considerable interest, contrasted with 49 percent in Britain, and 10 percent in Britain claimed they were not at all interested in new medical discoveries compared with 3 percent in the United States.

Media attention was caught by a quiz of scientific knowledge. A number of the questions were what the researchers called morale-booster questions—they felt that it was very likely that most people would get them right. For example, they asked whether it was true or false that hot air rises, and 97 percent of the population sampled in both the British and the American surveys got the answer correct. But there were a number of questions that many people "got wrong." For example, less than half the American sample and under a third of the British respondents knew that electrons were smaller than atoms. Similarly, only 63 percent of the British sample knew that the earth goes round the sun, rather than vice versa. Another question asked whether it was true or false that antibiotics killed viruses as well as bacteria. Here, only 29 percent of the British sample gave the correct answer and 26 percent of the American sample. In both

countries, younger people tended to know more than older people, males to know more than females, and middle-class people more than working-class people.

How much does all this matter? So what if only x percent mark the right box on a questionnaire? There are undoubtedly problems in interpreting what is going on when people answer questionnaires; for a start, posing true/false questions implies a right/wrong answer. In turn, if respondents consistently pick the "wrong" one, how do we interpret that? Have they just got it wrong, or do they have a good reason for selecting that answer? For example, 4 percent of the sample in the 1992 U.S. survey rejected the answer to a question about scientific method that involved a control group of people in a drug trial; they did so, it transpired, on the grounds that they thought it wrong to deny the drug to people who needed it.

Similarly, only about a quarter of respondents in the British survey knew that antibiotics are ineffective against viruses, but then, how many had the experience of doctors' prescribing antibiotics for them when they were sick with viral infections? Interestingly, women were more likely to get the answer right in recent U.S. surveys (reported in *Science Indicators*, 1992), perhaps as a result of their experiences of taking children to the doctor. Might women be more likely to question the need for antibiotics for their children?

Large-scale surveys have the advantage of (relative) statistical reliability, in that they draw on large numbers of people and provide a more representative sample than smaller scale studies. They can also provide a fairly coherent set of questions that could then be compared across different countries or cultures. Nevertheless, surveys do have many critics. A primary criticism of them has been that they perpetuate a deficit model, implying that "the public" lacks appropriate knowledge. The implication for science education is that people are empty vessels that need to be filled up with the right liquid information.

People are not, however, empty vessels; we accept or reject information and make sense of it by reference to context and to our past experiences. Other research suggests that people often have a complex understanding of science. "Anti-science" reaction occurs in particular contexts in which people have learned that "scientific expertise" is fallible or not to be trusted. Many feminists are highly critical of scientific research and medical practice because women have too often fallen foul of medical "experts." Antagonism to science is widespread in feminist communities for good reason.

In research into how people might perceive science, one study focused on sheep farmers in the northwest of England after the Chernobyl fallout (Wynne, 1991). For some time, there was a government ban on movement of the sheep from the hills. The researchers were concerned to find out how the farmers understood what was going on and their opinions of government statements about possible risk. Farmers were angry about what they saw as continuing delays in getting their stock to market; they were also skeptical about whether the government experts did in fact have adequate information on the levels of radioactive cesium in the grassland covering the lower fells.

The dominant view held by the British Ministry of Agriculture, suggested the researchers, assumed a deficit model: the farmers had failed to understand the message. But there was also failure on the part of government experts to understand the contexts in which the farmers were working. The officials did not appreciate that science can be parochial—the farmers might know more than the Ministry of Agriculture about the different rates of grass growth and the ways in which these grasses might take up radioactive cesium. Nor were official idealized measurements of radioactivity related to the way water collects in small ponds on the fells and which ponds sheep prefer to use for drinking. Government experts paid no heed to the specialized knowledge of the people; instead, they simply assumed that the sheep farmers did not know enough of the science.

Information is also something that people may choose to reject. The Cumbrian farmers, for example, refused to undergo whole body radioactivity scans on the grounds that, if doctors discovered high levels, nothing could be done about it. At the same time they were angry because their requests for water analysis were ignored—even though water content could be corrected. From the perspective of the sheep farmers, useless knowledge was offered but useful knowledge was denied.

Discussing such studies of how people react to scientific expertise, the researchers concluded that the main insight "is that the public uptake or otherwise of science is not based upon intellectual capability as much as social institutional factors having to do with social access, trust and negotiation, as opposed to imposed authority. When these motivational factors are positive people show a remarkable capability to assimilate and use science or other knowledge derived inter alia from science" (Wynne, 1991, p. 116).

The policy implications of an approach which stresses that people can reflect upon their relation to scientific knowledge are very different from those of the deficit, top-down model of the survey approach. Surveys have been informed by a particular rhetoric of encouraging and nurturing democratic participation. According to this, the level of public scientific literacy must be raised to that required to make informed judgements. Yet, as Wynne and his colleagues noted, the recommended ways of achieving this are based on the power relations of a dominant science and a subservient public. Thus, the Royal Society recommends that the amount and quality of science education should increase and that scientists should actively popularize it. Yet such actions are deeply imbued with power. They imply that the scientific community is the active holder and disseminator of knowledge; the public is merely the passive receiver. There is no room in this model for people actively to create their own understandings, nor to use these. If they do, what they create will not be called science.

Science, as we now understand it, is a particular way of knowing the world that grew up in the context of European expansion and the development of capitalism. It is also a worldview thoroughly grounded in gender and race (Merchant, 1980; Schiebinger, 1989; 1993; Harding, 1992). It has come to be seen as representing some ultimate truth, and in so doing, the practice of science has involved the denial of other ways of knowing the world. Scientists energetically dissociate themselves from what they believe to be "pseudoscience" (such as astrology, or even homeopathy). They also ignore the expertise of those people deemed to be non-scientists and of communities whose knowledges are not generated by those who are recognized as "scientists." In this way, the local knowledge of, say, the medically useful plants of the Amazon has been largely ignored. It does not count unless scientists have made—and claimed—the discovery.

The gap between the scientific expert and members of the public who are believed to be ignorant of or uninterested in science is a relatively recent phenomenon. During earlier centuries, some historians of science have argued, there was more general interest in science as another form of human knowledge. Scientific knowledge in Europe in the eighteenth and nineteenth centuries was disseminated in a number of ways, including the activities of the Mechanics' Institutes and the literary and philosophical societies. Travelling lecturers gave public demonstrations, and books, magazines, and pamphlets were produced in order to popularize science. Indeed, many of the women whom feminists have now found to be associated with the history of science were known for their popularization rather than for their work in original experiments. That their names are forgotten says much, both about gender and about the values placed on making science accessible to people.

Some historians have suggested that an important feature of this interest in popular science was the development of what has been called "science for specific social purposes." "Chemistry for precious metal prospectors in the 19th century was different from chemistry for agriculturalists and again from chemistry for public health officials" (Layton et al., 1986, p. 30). Scientific instruments, too, were part of the general culture, so that the *London Magazine* of 1828 could report that, "in every town, nay almost in every village, there are learned persons running to and fro with electric machines, galvanic troughs, retorts, crucibles and geologists' hammers" (ibid., p. 32).

What seems to have mattered was the question: what do we need to know for such-and-such a purpose? This approach to knowledge, rooted in need-to-know and experience, is quite different from the approach embedded in science education today, which is too often far removed from everyday life and whose "need-to-know" is related to technological control over nature. By the turn of the twentieth century, science as part of everyday life had begun to change, to develop into the distant institution it now seems. Science conceived as an apolitical, universal, empirical and uniquely objective form of knowledge came to dominate (see Stepan and Gilman, 1993).

Three factors contributed to this change: the first was the growing professionalization of science. The scientific community increasingly pursued its own interests, or those of particular individuals or groups in power (today, these are largely governments and funding agencies). A second, not unrelated, factor was the adoption by schools in England and Wales of a science curriculum marked by abstraction and apparent disconnection from the social values of the wider society. Third, this change in science education took place alongside a move to deradicalize the self-education practiced in many working-class communities in Britain and to replace it with "provided" education; that, in turn, was intended covertly to instil the values of the more powerful classes.

The Mechanics' Institutes, for instance, had begun by being centers of critical political education for working-class men (although some women also participated). But those interests changed in the 1820s through the 1830s, as local businessmen pushed for change. They saw the institutes as "potential centres for technological innovation and scientific discovery.

With their help they felt the Institutes could produce many more Watts and Stevensons from the ranks of the working class" (Cowburn, 1986, p. 110). Thus, the Institutes began to move away from their radical origins and to become incorporated into the values of those in power.

The emphasis on abstraction in the development of science education served the interests of particular groups in society. It taught a particular kind of science, divorced from people's needs and lives. And it has had a number of consequences. It encourages the rote learning of scientific facts, separated from any kind of context. Pupils must learn to reproduce some ideas in order to pass examinations but often feel that they do not really understand. It also makes learners become passive receivers of knowledge and teaches them to ignore or devalue their own cultural knowledge. Many of the women we interviewed recalled such experiences. The tales they told spoke of a science that had little meaning for their lives.

Abstraction is deeply embedded in the practice of most educational institutions. It is not unique to science although it is undoubtedly in science where it is most explicitly and highly valued. Abstraction creates conditions that marginalize many, particularly those who are poor and/or nonwhite. In analyzing reasons why young women from poor families drop out of high school, Michelle Fine and Nancie Zane (1991) point out that educational practice typically ignores private lives, where there may be particular difficulties or struggles. Many young women dropped out, they learned, because of family responsibilities—not because they felt these to be excessive but "because their schools imposed an artificial split between 'public' and 'private' issues." Life, as several students emphasized in their study, is rarely that simple. That gap between lived experience and what is taught in schools (and what that gap feels like to women) was a constant theme in our own research. Science is taught as decontextualized, and even if it is taught in ways that might connect it with the world around us, that is likely to be the public world (and a public world defined by a tiny elite), far removed from "private" life. Small wonder, then, that many women feel that science is irrelevant.

## Images of Science and Silenced Voices

As we have seen, women receive low ratings in surveys of public understanding of science. On the basis of those surveys we can say that women know very few of the facts of science. Perhaps that matters little; after all, many people in our culture (including most politicians!) do not know much science. On the other hand, among women who are seeking to return to education, there is often an interest in science combined with a feeling that "I couldn't do it." That combination—of potential interest and alienation—is noticeable and disturbing.

As a starting point for our research into how women outside professional science perceive it, we wanted to explore some of their images of "science" or "scientists"—what did these words mean to them? It is partly these images and beliefs about science that might stand in the way (or not) of women's access to scientific knowledge. In spite of the drawbacks of using questionnaires, we asked a number of women to write down some word associations. Word associations can serve as a kind of cognitive mapping, a means of sketching out the images that are evoked.

This open-ended questionnaire was circulated to 120 women belonging to a variety of community groups or adult education classes. We do not, of course, necessarily know the context in which individual women came to acquire the images and meanings evoked in response—but we would not necessarily know those contexts in the classroom either. We would, however, meet up with a variety of images and stereotypes about science.

The word associations, not surprisingly, tend to produce stereotypes. (One respondent complained that the question "asked for a stereotype"; indeed, that was what we were seeking.) For example, the "mad scientist" in a white coat—who could, for one respondent, terrify "me and my cat!"—was a frequent image. But being a stereotype does not make the image less relevant. If this persona also has gender (male) and race (white) then it clearly means "not me" to those who are women and/or nonwhite. It also means "not me" to those who see themselves as not "clever," since scientists were seen as particularly clever—if boring—people. "How can they hold all that stuff in their heads?" wondered Chris, in interview, as though learning involved only "minds," not bodies (see Foucault, 1979). Chris vividly portrays a view of scientific knowledge as exceedingly rational, cerebral, and "up there," disconnected from emotions, bodies, and "private" lives, "down here."

Scientists were considered too narrow-minded, unable to think clearly about the ethical and social consequences of their work. Some women emphasized the possibility of corruption or competitiveness, which could prevent a scientist from being responsible about his/her work. Most described research or experiments as "what scientists do," although some mentioned trial and error, or guesswork. One answer recognized the com-

bination of the tedium of research and its excitement by describing scientists' work as "slog and boredom—and euphoria."

In her analysis of images of scientists in western literature, Roslynn Haynes (1994) identifies six overlapping stereotypes that thread through several different genres—novels, plays, and films. These are: (1) the alchemist, who appears as the obsessed or maniacal scientist of the Faustian legend; (2) the stupid virtuoso, out of touch with the real world of social intercourse—the absent-minded professor; (3) the unfeeling scientist, suppressing emotion in pursuit of science— Dr. Frankenstein is a familiar example; (4) the heroic adventurer, such as scientist space travellers, exploring new territories; (5) the helpless scientist, out of control of the results of his/her experiments; (6) the idealist scientist, a believer in a scientific utopia. These images, especially the negative ones, recur throughout western culture, from Mary Shelley's gothic novel, *Frankenstein*, to the scientists and doctors of many of Nathaniel Hawthorne's short stories and the more recent Stanley Kubrick film, *Dr. Strangelove*.

These familiar stereotypes appeared in our questionnaire responses, as Table 1 shows. We have added a seventh category, clearly enunciated by many of the women, of the elitist scientist, who is motivated to maintain his/her place in social hierarchies.

It does not matter that the responses were, indeed, stereotypes. What matters is the persistence of these images and their general negativity. Those people who are outside science (and, in different ways, those of us who are inside it) inevitably absorb such images within the wider culture. And we bring them to our understandings of what it might mean to study science or simply to read a book about it. The scientist as heroic adventurer, for instance, is very apparent in the quotation from Bodmer and McKie's book with which we began this chapter. That image is critical to what philosopher Mary Midgley (1992) has called the modern myth of science as salvation—the idea that science, in particular, can solve humanity's ills.

The two more positive characterizations in Hayne's typography—the scientist as idealist and as hero—were relatively scarce in our questionnaire returns. In some ways, this is perhaps unsurprising: the image of heroism, at least, is imbued with gender, and the scientist as hero is a figure from which women are likely to feel distanced. The negative images of scientists, as different, peculiar, obsessed, or dangerous, prevailed.

Science may contribute to progress (and many respondents did allude to that), but a more consistent theme across questionnaire returns was the difficulty and potential danger of science. It was also described as "inhuman," cruel, inflexible, and unreliable. We expected negativity in responses to starters such as, "The trouble with science is . . ." but even the more positively phrased beginnings, such as, "The purpose of science is . . ." or "My image of science is . . ." elicited negative descriptions from some of the women.

TABLE 1

**Stereotypes of Scientists** (after Haynes, 1994)

**1. The Alchemist**

" . . . my image of a scientist is someone who destroys nature"

" . . . someone who wants facts and answers to life, the universe and everything . . .single-minded, often dreamers and idealists who long for stardom by discovery (like finding the Unifying Force) and thereby gaining a form of immortality"

" . . . the end product of research becomes a sort of Holy Grail that they must keep questing after no matter what its effect on society"

**2. The Stupid Virtuoso**

" . . . loveable but a social outcast"; "no personality"

" . . . they produce a one-sided outlook on matters . . . because they are men"

"they affect family life because they have to check and observe the processes of the experiment"

"a man with long hair and a beard and spectacles . . . locks himself away doing equations and measurements. He doesn't relate well to ordinary persons but is very clever and dedicated"

**3. Unfeeling**

" . . . a white male, locked away in labs, experimenting on animals—ugh!"

"irresponsible and uninvolved in the broader implications and consequences of their work, either through lack of opportunity to express them or encouragement to do so"

"do not see people as human beings—or animals as living, feeling creatures"

**4. Heroic**

"dedicated to a better future"; "constantly striving"

**5. Helpless, out of control**

" . . . they have gone beyond a boundary—for example, in vitro fertilization."

"they never find solutions to the mess they leave behind"

" . . . can go beyond the limits of what is acceptable and call it progress"

**6. Idealist**

" . . . a mission to give people many benefits"

**7. Elitist [additional category]**

" . . . speaks in a language that alienates everyone who is not a scientist—elitist"

"speaks a different language"

" . . . they think because they understand and find it interesting then everyone else should"

" . . . danger of believing they have an absolute knowledge or the power to find it and the conceit that goes with that belief"

For many of those whom we later interviewed, these images of science and scientists contrast with a perception of their own (or more generally, women's) understanding(s) of the world. This was seen as more "common sense," rooted in everyday experience, than science could ever be. For some women, this was expressed as a personal, owned understanding; others contrasted science with an understanding that they had deriving from particular histories. Black women, for instance, spoke of their cultural understandings in opposition to white culture—for example, traditional knowledge of herbal medicines in African countries compared with scientific medicine.

Despite that contrast between science and common sense, science and its language does influence the everyday language of all of us. Emily Martin's research is instructive here. In her 1989 book, *The Woman in the Body*, she noted the metaphor of the body as production system that predominates in medical textbooks. This metaphor constructs female physiology as failed production—menstruation as the breakdown of the endometrium or menopause as the breakdown or ending of ovulation.

Yet women do not always make reference to that metaphor. In asking women how they would explain to a young girl what happens to a woman's body at menarche, Martin found striking social class differences. Middle-class women gave an explanation grounded in a scientific model: they referred to changes in hormone levels, to the breakdown of the lining of the womb. Working-class women used a phenomenological account that described menarche as a rite of passage to becoming a woman. Martin emphasizes that these women knew about the medical model. It was not a case of not knowing. It is a case, she believes, of an explicit resistance to the medical model and its implied failure and breakdown.

Martin's later work (1994) focuses on discourses of the immune system. She documents the richly nuanced language and metaphors used to describe the immune system in its dealings with infectious disease. But, she points out, the languages of scientists, members of the lay public, and practitioners of alternative health are all fairly close to each other; all emphasize the complexity of the immune system and its flexibility. What is striking in her account is the way that lay people "readily and vividly convey their sense that the immune system is a complex system in interaction with other complex systems . . . that changes constantly in order to produce the specific things necessary to meet every challenge" (p. 80). It is part of their commonsense notions of how bodies work: they had picked up some of the language, the ways in which immunity is now described in medical and scientific texts, so that their own, "lay," accounts mirrored those of scientific practitioners in many ways.

The women we interviewed have, no doubt, incorporated many such scientific ideas and narratives into their own worldview. They could probably have described the immune system (had we asked about it) in much the same way as the people in Baltimore described by Emily Martin. But most of the women were at pains to differentiate their own knowledge and common sense from what they saw as "science." However much

information and discourse they had in fact taken up, they were likely to rename it as not being science—"if I understand it, it can't be science."

The power to name what counts as science does not belong to women; naming what counts is continually being renegotiated among those in power. The boundaries of science are constantly policed, both in terms of what counts as science and of what counts as good, or valid, experimental results. Much recent work in social studies of science has focused on ways in which scientists use language to persuade others, usually scientists, of the veracity of their claims (Collins, 1985; Latour, 1987). In doing so, they create or perpetuate boundaries between "real" science and what is not to count as science (parapsychology, for example). These boundaries are thus not fixed, but socially contingent: what counts as beyond the boundaries, as pseudoscience today, might be cast as real science tomorrow (Wallis, 1985).

Feminists have often emphasized that naming is a political act. The very term *woman*, for example, has historically had multiple meanings and contexts (Riley, 1988), and acts of naming have played a critical part in the colonial discourses that describe the "Third World" (Chow, 1989). The women we interviewed named their own experience and knowledge as owned, personal, and individual. Science was an "other" kind of knowledge, belonging to social groups of which they were not part, and their naming of it in this way reflected that separation. What they knew, they emphasized, was "just common sense."

What counts as common sense, of course, depends upon one's vantage point. Wendy Luttrell's account (1989) of working-class women's use of concepts of common sense as their own knowledge contrasts that knowledge with the "stuff that comes out of books." Yet that kind of "common sense," or everyday knowledge rooted in working-class culture, may differ between people; in our study, black women's explanations made links with African cultures or those of the West Indies. So, too, might the "commonsense" knowledge of women in poorer, subsistence economies differ considerably from that of white working-class women in Britain.

The concept of "common sense" is partly universalizing, but it is also partly individualizing and oppositional. It stands in opposition to what has been called the "master narrative" of science (Lyotard, 1984). Science has a voice of authority; appeals to scientific proof tend to win arguments and give power to those who can articulate them.

Not everyone has access to political processes and the power of that authority. Understanding science well enough to take part in debates about, say, the effects of local industry on the environment, may be desirable but is in practice likely to be the prerogative of only a minority, even in richer, industrialized nations. Who will have access to the knowledge? Whose opinion will be likely to count in debates? Whose opinion will influence politicians? And how will those opinions be heard? What we can be sure of is that an opinion based on perceived common sense will hold less weight than one based on "scientific evidence." We can be sure, too, that the passionate eloquence of a woman who has noticed that environmental pollution is making her children sick (however much science she has learned) is not likely to be heard—especially if she is neither wealthy nor white nor living in the industrialized West.

Not only is she unlikely to be heard, but the complaints of workers that the workplace is making them sick are also usually dismissed on the grounds of a lack of "scientific evidence." Scientists (and politicians) ignore mounting "circumstantial" evidence that people in a particular industry or workplace are becoming ill, trusting that epidemiologists will do "properly controlled" and "rigorous" studies that will establish whether or not there is a link. That these are often biased (leaving out women, for instance), sometimes use inappropriate statistics, and ignore the statements and experiences of workers themselves should be grounds enough for doubting their credibility. What counts as "acceptable risk" is something on which supposed experts fail to agree. The concept of risk is evaluated differently by different government agencies. For example, in the United States, federal agencies such as the Nuclear Regulatory Commission, the Food and Drug Administration, and the Environmental Protection Agency use dramatically different criteria for assessing public exposure to chemical or radioactive hazards (Block, 1994; see also Grossman, 1993).

Yet such criteria are typically lauded as "objective" and "scientific," providing the only "proper proof" of any connections. That they are often used by industrialists and governments to "prove" a lack of connection between dangerous chemicals and ill health should be evidence that they are anything but "objective." Feminists Donna Mergler and Karen Messing (1995), writing about statistics in occupational health, have noted this.

The fact that some people's voices will not always be heard does not mean that they should, or could, be silenced. Indeed, as Messing and

Mergler have noted, workers are (relatively) lucky in that they are not as helpless as laboratory rats: they can (sometimes) protest the inhumanity of science, at least if they are workers in the West.

One way of silencing is to deny access to particular forms of knowledge. In the case of workers affected by dangerous chemicals, knowledge of what the chemicals might do may be denied by the employing industry and that denial colluded with by both politicians and scientists. The knowledge is made inaccessible through the specialized language and abstraction of scientific writing. That inaccessibility is greater still if the workers are in less industrialized countries. The science is written in a language that is not theirs (English, say), and western companies can evade responsibility by blaming local operators/laws. This happened in the case of the disaster at the Union Carbide chemical plant in Bhopal, India.

The opposition between scientific knowledge and "what people know" is not a symmetrical one. A mother may believe that the chemical plant down the road is poisoning her children; she may look at maps showing clustering of symptoms around the location. But without the backing of scientific arguments, her common sense will achieve little. Scientific experts will attempt to disprove her claims. They will assert that there are too few cases, or that the clustering could have arisen by chance.

The controversy surrounding scientific literacy/public understanding of science fails to address the ways in which people may categorize knowledge. Knowledge that is useful to them in their everyday lives is likely to become named as common sense, while "science" is something unreal, divorced from reality. Knowledge may actively be used against people, in the ways we have just described in relation to environmental pollutants. The point of the more qualitative research on public understanding is that it shows that people do indeed find out about, and use, scientific information when they want to for particular purposes. It can, in those contexts, become part of their everyday understanding of the world.

Although we may welcome the great efforts being made now to promote science, we should do so with reservations. Having fun with science, producing plays about science, mixing art with science—all have an important part to play. Yet these efforts still tend to reinforce the notion that science is the prerogative of "special" people, that it is far removed from everyday lives. Partly, this is specific to science and its voice of authority; but partly, it is a response born of years of experiencing an educational and social system that denies agency to many people. In that system, many forms of knowledge become removed from everyday life. Thus, "I don't understand all this politics stuff, it's not for the likes of me," commented one woman in a study of working-class women's experiences of involvement in women's groups (Butler and Wintram, 1991).

Promoting a "gee, ain't science wonderful" response without encouraging people to feel that they, too, can be part of it is likely only to perpetuate the boundaries between those who have, or have not, got what it takes to be a scientist. As we shall see in the next chapter, the belief that "I don't have what it takes" is a powerful reason for girls in high school to avoid math or science, which they see as something "only really clever" people can do.

# *Case Studies: Cultures and Contexts of Adult Education*

In this chapter we move from standard chapter format to present a series of case studies from our research interviews on women's perceptions of science. Overall, these interviews differ in that they reflect varied life histories related to age, class, and ethnicity, and one noteworthy difference is that between women from different groups or institutions, illustrated in this chapter. The case studies given here are not atypical, but we do not claim them to be in any sense representative. They stand on their own, as vignettes. To begin with, we have selected outline case studies of two individual women to provide examples. In chapters 4 and

5 we shall move in some detail to the findings of our research interviews.

Some women we interviewed were attending courses, others were part of community groups. As we interviewed several women from the same group/course, we noticed consistent themes emerging from within each social group. There was also a marked tendency for women from the same group to answer in similar ways.

The belief, for example, that science means "lots of studying" by "clever people" came from half the women in one group (of six interviewed), while in another group that was involved in literary and cultural studies there was a greater attendance to issues around stereotyping than in any other group—or across groups. This emerged most clearly in their questionnaire returns and took one of two forms. It involved either a refusal to attempt to pin down science or scientists, for example, "a very vague area because it is all-encompassing—the name science is very ambiguous" and "this asks for a stereotype," or it gave a self-consciously stereotypical answer, such as a scientist as "a stereotypical/boring/spotty/anti-social/ugly student-type person in a white coat i.e. Brains from Thunderbird."

In another institution, one renowned for its emphasis on experiential learning and for its women-centered approach, half of those interviewed shared a distrust of experts, a view that science should "come out of the closet" and a refusal of the notion of "scientific facts." Such refusal typically construed scientific facts as "right or wrong, black or white" and seemed to indicate resistance to any notion of objective knowledge. As one woman put it: "Scientific knowledge is proven knowledge but in the end it's personal choice what you believe." These women were particularly aware of the politics of science and its place in society.

We examine this latter institution in more detail here, contrasting it with another women-only group that is studying science with feminist tutors. We do so in order to highlight how narratives as ways of understanding the world are developed in different contexts. Our focus here is on the stories about science that people outside science construct and how these may be influenced by their social milieux and interests. In looking at the different groups, we are interested in how they may produce different epistemological communities in their discourses around science.

The third case study is different. It is based on a group of Muslim women, living in Coventry, rather than on a group based in a specific college or course. These women share the foundation of Islam, which provides them with a framework within which to interpret science and also

the place of women in society. Part of this case study is based on a group discussion with ten women, in which we raise some specific issues about genetics.

## Individual Case Studies

### Alice

Alice is a white woman in her fifties. Her high school education took place partly in Britain and partly in Australia. On returning to Britain in her midteens, she went out to work because of family finances and continued her education through classes in the evenings. Now she is studying literary and cultural studies, with the hope of taking a degree course. Alice has always been interested in literature.

Despite an interest in science, she did not continue it beyond age thirteen. This was determined by gender biases—"the boys did science, you see," she said.

Alice sees science as a form of knowledge that differs from others in that we cannot reject what has already been done (perhaps for that reason, she also expressed concern about the need for controls over genetic engineering). Speaking of religious knowledge, by contrast, she feels that "in the end, you either accept it or reject it, but with science you can reject its findings, but you can't reject what it's actually done." What differentiated them, she believes, is the facts of science, the accumulation of evidence.

Later, however, she contextualizes this by noting that how people interpret what scientists say also matters. Here, she is talking about the fashion for "natural health," especially related to food. She suggests that people might be more willing to take risks with food; they might, she says, "eat the things that are bad for us and are proving to have chemical reactions in the body [but accept it] because it's food, but if it wasn't a food and somebody said that was bad for you, science has proved that it's bad, you wouldn't touch it with a barge pole, but because it's food you tend to think, oh, it can't do any harm."

Like many other women we interviewed, Alice relies on a "mixed" account of childbirth—using medical and phenomenological terms, emphasizing the experience for women. Thus, she explains, "if I had, say, a daughter today I would explain probably in the same way my mother did, and go through the reproductive system, but I think I would try to explain a little bit more about how it actually feels when you give birth and the importance of it, and the feeling that even if it hurts a bit it's worth it in the end."

Alice spoke about the seasons and nature, invoking the myth of Persephone and the underworld, talking about the "world's reproductive cycle . . . spring is the birth of everything. The beginning, the fresh beginning, summer is the nurturing, autumn is getting ready for death, and the winter is dying and left to be reborn the following spring" (see also chapter 6).

Alice also illustrates the importance of understanding how things are learned. She had learned about childbirth partly from her mother and partly through experience. Learning how to use technology, such as a microwave oven, came from reading about its potential hazards. In her explanation of the seasons, Alice returns to the social context of learning, specifically through women she has known:

> *. . . observations of nature and through people who I've known. . . . We have two old ladies who live next door to us . . . they are marvellous. My own grandmother was 93, and talking to them, you observe their lives have been through the same sort of cycle as a plant or a flower. You know, I don't see death as the end I see it as part of the process. You know, as a flower dies and the seeds drop and it comes again and it's all part of the cycle.*

Not surprisingly, perhaps, Alice felt that women would bring something different to science, "because a woman feels things differently, a woman's body is slightly different to men's bodies so probably what she would do . . . might be different." Women, she feels, are more likely to consider the consequences, "if only for the fact that they are involved far more in the reproductive cycle than the men so therefore their concerns for mankind are different from a man's concerns."

### Bell

Bell is a black woman, aged twenty-six, with a six-year-old daughter. She left school at sixteen, then did several jobs before hearing about Hillcroft College. She now studies various subjects, including women's studies, there and hopes to go on to university. She chose Hillcroft partly because it was a women-only college and partly because it was specifically for adults, which she felt would be less intimidating.

To Bell, scientists are "in a little world of their own." "I don't think of the scientist as being like on a par with anyone else . . . I think there's a

wide gap between them and us." Partly this is because she sees scientists as clever, but also partly because science itself is "out of reach . . . it just doesn't seem real, it's the stuff novels are made of, do you know what I mean?"

On being asked what she thinks about science, Bell feels that, for women, getting through a day is a major achievement, hence scientific: she notes that "it embodies so much more than just, you know, Einstein's theory of relativity or whatever, I don't know, I think just getting through a day on the earth is like a science on its own. . . . It's like a part of this massive game . . . like this massive chess board or whatever and you have to make the right moves in order to get on and there's so many rules that we have got to abide by. . . . It's a miracle for me like especially if you are women of this world if you get up like 7 o'clock and you get through and you go to bed at 11 . . . and you still feel OK that's a major achievement."

After talking about having to face the real world after high school, and how society ascribes certain roles to women, Bell goes on to say that "I don't feel that there is anywhere a woman can go to be taught about science or whatever."

Science to Bell is in everything; yet at the same time she says she "didn't do anything scientific" (although she goes on to admit that she lives alone and has to deal with household technologies). To her, science is "about things you can feel and you can touch," about things that are "real" as opposed to belief.

Bell resists categories of scientific versus nonscientific, even in relation to a list of academic subjects. History is scientific, "not the sort of stuff about battles and that but the sort of history we study here is verging on the scientific because it's more about people and the ways of life rather than great men and women and battles and stuff like that you know."

Like several other black women, Bell notes that women in her family had a folk tradition of making and using their own herbal medicines which they often relied on rather than approaching doctors. Conventional drugs, along with food additives, she considers to be going against nature. It is around food scares that she expresses (in contradictory ways) a mistrust of scientific experts:

*I only found out the other day . . . that we are supposed to have the worst meat in Europe . . . and I'm thinking, God, what have I put inside me all these years . . . and it's like I trust because you know we live in this country and we got the Government that we have got and you sort of like expect that they are out there and we trust what they are doing so you go into a shop and buy the meat . . . and you go home and give it to people you love and care about—you are probably poisoning them and you don't even know. . . . but a lot of what we get [information] . . . is probably, like, I don't know, blown up out of proportion or whatever because you know you hear about this story and I know it's real and then you get some people running around and saying . . . watch what you are eating . . . and making it bigger than what it actually is . . . if you stopped and took a deep breath and really thought about it whoever it is that does could control it, study it, and find out why it is happening and you know under what sort of circumstances . . . and deal with it. I mean it gets out of hand when people are saying that two people have died of this or that . . . because people start to worry so I think if they just realized science could find out things . . . .*

Speaking of childbirth, Bell produces an articulate account based on a scientific story. She is much more hesitant when talking about microwaves. Her account of the seasons begins by resisting the scientific story and moving on immediately to her memory of a Japanese version of the Persephone myth.

Bell feels that we would benefit from having more women in science because women have more compassion and empathy. If in turn science had more empathy, she thinks, it would make it "more acceptable to more people, it opens it up a bit . . . you bring it down to a much more human level. . . . [as it is] it's cold and it's clinical [but] especially younger children have got to understand it's not a boring subject . . . [but] we have to sort of make it get down to a much more personal level and it's not there [but] if we had more women there . . . women are more in touch with their emotions."

She is also concerned that she knows of no black scientists, male or female, and points out that "at school you don't talk about the contribution black people have made."

### GROUP CASE STUDY I

Our first case study centers on Hillcroft College, a residential college for women lying south of London. Residential adult education is central to the distinct tradition of adult education in Britain, and Hillcroft, as the only residential college for women, is a microcosm of changes in that tradition. Its changing emphases also reflect (and contribute toward) changes within feminist thinking about women's education.

A shift to access to higher education courses (courses designed for

adults who lack traditional entry qualifications for university) in recent years has not compromised Hillcroft's renowned emphases on a women-centered approach and experiential learning. Importance is placed on developing women's confidence in their abilities as learners, drawing on their life experiences and strengths, and stressing the emotional as well as cognitive dimensions of learning. Science is not part of the curriculum, although staff would like it to be included. They regard learning about science, its relevance to society and environment, as integral to learning in science.

In the Hillcroft interviews there is a stress on personal knowledge and on what Belenky et al. (1986) call "connected knowing." Connected knowing values truth that is grounded in firsthand experience and validated through shared experience (Belenky et al., p. 118). Empathic understanding is its key feature. What has been called a "counselling discourse" permeates the Hillcroft interviews (see Fairclough, 1989; Kitzinger and Perkins, 1993) and a notion of "blocking out" science and scientific knowledge is peculiar to some of them. For example, one woman, voicing the attraction felt by all interviewees to Hillcroft's caring approach to its students, says, "It's opening things up, helping me explore myself." Another student, who finds it difficult to speak in large groups, comments on her philosophy course: "I have the same difficulty with philosophy as I'd have with science, I think. I can't get into the argumentative mode required." (Michele Le Doeuff [1991] has suggested, on the basis of her own experience as a philosopher in a male world, that if you are a woman and a philosopher, it helps to be a bit bad-tempered!).

Similarly, Chris, a woman who in her questionnaire return, expressed an oddly self-conscious view of science as "a subject I'd rather stay away from. Science means not me. Science means asking why?" continues the theme in her interview: "Other than the psychological side of things I'm quite ignorant of anything 'scientific.' But I'm quite proud of that. Not quite glad that I'm ignorant, but it doesn't bother me at all. My brain has just shut off from it completely." This comment, a mixture of defensiveness and pride, sums up well the kind of resistance to science—blocking it out—shared by half of those interviewed. At the same time it implies that it is "connected knowing" that Chris feels she possesses.

When asked to recount their most significant learning experience, all refer to experiential learning that is personal or relational—events such as giving birth, watching a child grow, understanding anger and forgiveness for the first time, being a drug addict and waking up to cherry blossom, identifying with nature, a leaf, even: "just touching a tree and feeling vibes through my body. Feeling at one with nature."

There is awareness of and distaste expressed for what Freire calls "banking" education (Freire, 1970) and a view that science education epitomizes this: "It's having so much contained in your head of that kind of knowledge." Hillcroft's experiential approach to education is strongly preferred. Anger at the oppressiveness of an educational system that leaves a person in the dark is apparent in one woman's comment about science at school: "Personally, I can't relate to the whole idea now. I found that teachers never explained what the hell was going on. Sticking things in test tubes . . . I never understood the purpose of what I was doing."

Skepticism about expert scientific knowledge and methods for acquiring it was expressed in the following discussion with the focus group. In these groups students were asked to read and comment on a number of mock press reports on scientific issues (including one on arthritis research). The extract also illustrates an aspect of what has been called the "condition of modernity" (Giddens, 1990; see also Beck, 1992), that is, the lack of control or sense of powerlessness that many people feel in a world increasingly dominated by technical and scientific experts. Expert information, as well as recycled knowledge (via the media and friends, for example), is often inconsistent, and yet, in order to act, we must believe in something.

This extract is taken from the middle of our discussion.

*Interviewer: Listening to you, it sounds like you're pretty skeptical about what you read in the newspapers.*

*Student A: The trouble is you read something one day and a few weeks later something that contradicts it, so you've got to be skeptical.*

*Interviewer: So if you needed to find out something, say to do with your health, what would you do?*

*Student B: Read books.*

*Student C: If I had arthritis diagnosed I'd not rely on written evidence. I'd talk to other sufferers because I'm skeptical about anything I read.*

*Student D: But you've got to believe somebody.*

*Interviewer: I think C is suggesting you can't always rely on expert opinion.*

*Student D: I've a horrible tendency to believe them even if I don't like them or what's said. If it's got a touch of the scientific I'll believe it.*

*Interviewer: So it's about who says it?*

*Student A: There was an article about the need to eat more meat . . . and when you looked closely it was research paid for by the Meat Council. So you need to look at who is funding the scientists.*

*Student D: But it makes me feel inadequate and who am I to say this is rubbish when I don't know enough about it?*

*Student E: I worked in a psychiatric unit and every year students have to do the same test on rats to prove the same thing. Computer data could have been used but no one said why do this? The students were just as moronic as the professors. And they killed these rats by bashing their heads against the table.*

*Student B: Maybe it's to do with learning from experience?*

*Student E: From when we go to school we're not taught to question anything and a lot of medicine is just tradition and they're trying to relate rats' brains to humans'. You'd think by now they'd have given that up, figured out you can't connect the two.*

Skepticism toward experts was clearly combined with wondering why "they haven't given it up" yet, alongside a recognition that "you've got to believe somebody."

The interventionist, controlling power of science is a dominant discourse in the Hillcroft interviews. One student remarks: "Nature is as Nature is and it's just constantly changing all the time but they [scientists] are just trying to speed up the process. Science tries to control and change things. It has created a lot of monsters and without science we won't be able to undo the monsters." Another woman, troubled by recent media coverage of research on "homosexual brains" and worried about the inaccessibility of expert knowledge, expresses the double meaning of manipulation when she says, "That's what's frightening about not understanding science. You think, how the hell can they do that? And you've just got to accept it. It could mean tampering with genes. It could be a load of rubbish and we could really be led up the garden path on some issues and that's really frightening."

We live increasingly, as Beck (1992) has recently pointed out, in a "risk society" that presents particular challenges to adult education. The women we interviewed at Hillcroft and from other groups are well aware of the ecological risks to which they and their children are exposed—through nuclear hazards, genetic engineering, and food additives. The social changes of the risk society, Beck suggests, help create a form of individualization such that people's lifestyles, in part, contribute to particular risks.

Yet in the risk society, individuals' experience becomes marked by the sense of powerlessness to which Giddens (1990) has referred. Adult education has devoted much attention in recent years to the issue of experiential learning. At Hillcroft and in women's education more generally it is regarded as perhaps the most important aspect of an education that can be empowering. But there is a danger that, in focusing on experiential learning, educators ignore the fact that many of the risks to which people are exposed are beyond daily experience. We cannot know through experience, for instance, how the ozone layer has developed a hole. In their ambivalence toward experts and the authority of science, the women are articulating this sense of science as outside our daily lives.

An empowering science education for them would have to make connections between lived experience and structures and processes not available within that everyday experience. In defining science and argumentative modes of reasoning as "not me," in valuing personal and connected ways of knowing over the kind of knowledge they see science as representing, they may deny their own capacity for knowledge that goes beyond the familiar. The women are also articulating a sense of the poverty of a scientific "rationality" which, in its narrow instrumentalism, subordinates other human needs to the goals of "efficiency" and profit. Their disdain for "proven" scientific facts, their preference for knowledge they feel they "own," and their dismay at the interventionist, controlling power of modern science, demonstrate this.

It is central to our argument in this book that paying attention to such resistance to knowledge, rather than ignoring or trying to "correct" it, is a way of creating new conditions for knowledge. It is a way of extending the boundaries of possible knowledge at any given moment and in any given social and cultural context (Nelson, 1993).

## GROUP CASE STUDY II

Our second case study focuses on a group of women in a science course for women only taught in Coventry. The Women and Science course (WS) has been running for four years. Initiated by a group of feminist scientists, it takes place in local schools in Coventry, an industrial town near Birmingham with a high unemployment rate and a large Asian population. The course, which has nursery facilities available, runs for one morning a week over six terms.

The course is part of Warwick University's Continuing Education "Certificate" program, which gives access to its undergraduate degree program. Participants need not complete the Certificate. Some women enter for one year only, using it as a step back into education; others want to leave their options open. WS is one of the few examples of a broadly based science course for adults that has no clearly defined vocational purpose. It is a rare example of a course based on science and taught specifically to women by women (indeed, by feminists). It provides a useful case study to explore our theme of epistemological communities outside science.

The course developed from a precursor called Inside Science (IS) that was held at a local primary school (Birke, 1992). This was taught around the examination of a series of specific topics, such as in vitro fertilization, pollution, and nuclear energy. The WS course was intended to be a much deeper, and broader, introduction to science, aimed at reducing women's fears of it and helping them to recognize what they already knew.

An important aspect, however, of such courses is that women often do not feel able even to come to the first meeting. It was crucial to visit community groups to encourage women, telling them how exciting the course would be, and how they could do it (Birke and Dunlop, 1993). Even when they arrive, some women are mistrustful, as they have "been told before that a course needs no qualifications, then when we get there we find you do need them." They have to be persuaded that tutors mean what they say—no prior qualifications nor experience of science are necessary.

A case study of the pilot course, "Inside Science," involved interviewing fifteen women who had attended all or part of an IS course (see Birke, 1992). Our later study of different adult education groups included interviews with five women who were at an early stage in their Women and Science course, toward the end of their first term. Similarities clearly emerge from both sets of interviews. Almost half mention that the course was for women only, some noting that at school they had seen science (except biology) as a boy's subject; others feel less intimidated, "more equal," without men, a finding that accords with other studies of women returning to study (see Thompson, 1983).

In the time-honored adult education tradition, the course starts from what the women already know and builds from there. It also focuses from the beginning on how other kinds of knowledge have been excluded—women's, for example, or the contribution of Islam to modern science and technology. This can be empowering, particularly when the subject is science. One woman, Jenny, who thinks it is important that women are teaching the course, explains: "The majority of men are rigid. I get the feeling that what we've done, you know, about nuclear energy, would have been very stereotyped. I do find a lot of men don't expect women to know and if you ask a question it's taken as questioning their ability to put over information. Another, Catherine, puts the point nicely: "They [the teachers] get at you from the inside. They bring out to the surface what you know."

All the women interviewed for both studies have clear perceptions of women's exclusion from science, beginning with school science. Those who had done some science at school refer to their sense of fear of things scientific. One speaks of having a "panic attack" in the middle of the session on atoms and molecules and "feeling how I'd felt at school—lost, almost" (9; Inside Science). One way of expressing this fear is to define science as whatever you don't understand. A woman who feels like this about medicine said, "I don't know enough about it, so it must be science!" (4; Inside Science).

Women in both groups claimed that before their adult education course they perceived science as highly abstract. It also has its esoteric apparatus, as one woman remarks: "Science is about life but before Inside Science I'd have said it's about labs, bunsen burners, test tubes, that sort of thing. Now I feel it's something that affects us all" (2; Inside Science). Another woman, Carla, who sees "real science, the most scientific" as "delving very deep, going really very deep into a subject, to the most basic, right down as far as you can possibly go . . . the hidden bit of things you don't see" thinks "the less familiar the words, the more scientific."

While the interviews reveal a distinction between science and common sense, they also reveal that one way in which the women's perceptions have been changed by the course is to include more everyday aspects in the category of what counts as science. The arcane world of the laboratory bench is replaced with "the world around us." This dominant motif of the

interviews is almost revelatory: scientific knowledge is not separate from everyday life. I do have scientific knowledge in my common sense! For example, "Before, it was test tubes and mad scientists; at school, physics was above my head. Here, it's areas that affect you. I didn't realize that pollution came under science . . . and everyday things like cooking. I just think that everything is scientific now" (Sandy). Again, "I used to think things like chemistry, things I don't really understand, and science labs, experiments [that you] didn't understand, experiments that didn't really work . . . I didn't really think of science as to do with the world around us" (10; Inside Science). The women in this course share a sense of science permeating everyday life, as not set apart from their own experiences; they share, too, a sense that what they already know, as "commonsense" matters.

Many of the women relegate some of the material discussed in the course to the category "common sense" and so not science. What is labelled as common sense are things that "you could discover for yourself," as opposed to the "very deep . . . things you can't see" (Carla). The contrast between common sense as "things we know" and science as things we do not is summed up clearly by one woman:

> *Well atoms and electrons, radiation, nuclear power—probably lasers [are all scientific]. It depends how you are actually looking at them. They all need science, but like the road building [topic], energy, it's more common sense . . . I suppose it's because we think we know a bit about those. Well, the rest, if you know nothing about it, it's totally scientific. I think it's something that's completely outside of your sphere, if you like, it's like going to the moon . . . it's totally scientific. (7; Inside Science).*

But what is significant from the point of view of "empowerment" is that it is precisely in these areas of "common sense" that they can identify things they already know. Nutrition, reproduction, and pollution are mentioned by several. One comments:

> *I think a lot of the [topics] you knew something which you didn't think you did . . . like the nutrition, a lot of the aspects that M covered you had a basic knowledge of that. And reproduction and I suppose radiation, things you'd picked up as general knowledge, that stimulated me really. It's stored at the back of the brain and you get a bit of stimulation and you think, oh yes, I know that. You had a basic awareness at least. (11; Inside Science)*

And one woman, Jenny, from the other group, when asked to say how she would describe scientific knowledge, says:

*"Mm, hard one, I mean what is knowledge? From my own experience I can know something without knowing I know something about it . . . a sort of inner knowledge and it's like bringing it into focus. [Scientific knowledge, commonsense knowledge] are the same, just a difference of degree."*

This comes close to the philosopher Quine's notion of knowledge as "seamless" and "all of a piece"—scientists' theorizing is no different from that of laypersons or even philosophers. For Quine, science in the broadest sense consists of almost all our efforts to organize our experiences. It is virtually without boundaries. Science in the narrow sense is just refined or "self-conscious common sense" (see Nelson, 1990, p. 109).

We could say, then, that a notion of science as refined common sense is beginning to appear in these women's discourse around science. It is, indeed, fostered by the course. Such an attitude toward science certainly creates opportunities for new learning in and about science; it may also foster (may even be a condition of) the production of self-critical scientific knowledge. Feminists and others have pointed out that the prevailing view of science (what Harding refers to as science's own "spontaneous consciousness") is that it is detached—that values and politics are irrelevant to the knowledge produced in it. This "lie" both maintains science's authority and insulates it from critical discussion of the values and social and political experiences (including those concerning gender) that find their way into scientific research.

Given the authority of science in our society, the need for critical self-reflection is compelling. But such self-reflection has to be done by communities—and not only scientific ones. Wrestling with questions of value and purpose is a matter for the larger community and should not be left to the few who are committed to the view that science has nothing to do with values and politics (see Nelson, 1990). By challenging the notion of science as something apart and separate from women's own commonsense knowledge, courses like WS open up a space for such scrutiny, while not absolving women's own knowledge from similar scrutiny. In so doing they create new conditions for the construction of knowledge and, potentially, enlarge for all of us what it is possible to know in any given historical, social, and cultural context.

## GROUP CASE STUDY III: ISLAM

The quotations we use here are from a discussion with a group of ten Muslim women in Coventry with whom we explored some of the issues about women and science that touched on their lives as Muslims. Some of these were British Asian women; others were women who had converted to Islam from Jewish or Christian backgrounds. Although there were important differences between them, what they shared was Islam: both in the group discussion and in individual interviews, their views were expressed *as* Muslim views.

Much of what is said is said also by other women we interviewed, for example, a belief that women and men "offer a different side to science," or that scientists are "going too far" in using animals to test cosmetics. They felt that science education (and particularly school science education) is "too dry" and irrelevant to their lives. As one of the younger women points out, for her, science would be fun and easy to learn "if it was taught through rap." As teachers, we weren't too sure of our abilities to teach through rap, but it is certainly an interesting idea to catch the interest of young people!

Many other issues came up that illustrated particular perspectives derived from the women's experience of Islam, as well as their experience as Muslim women living in Britain. For instance, several women agreed that Islam placed much emphasis on the need for both women and men to seek knowledge, including scientific knowledge (also see Afzal-ur-Rahman, 1981). One woman felt that Muslim women "have so much to look after," including knowledge of what food is permitted or forbidden, "that we have to know about science." Most of them were well aware of the significant role played by Islamic culture in the development and maintenance of learning science (indeed, many Arabic words have found their way into the history of science, such as alchemy, algorithm, alcohol, and algebra). On the whole, however, they did not learn these connections in school.

For Muslims, they emphasized, it is particularly important to learn about science in order to learn about their religion. Here, they were implicitly appealing to what Leila Ahmed has called the "ethical, egalitarian voice" in Islam, contrasted with the legalistic voice of many establishment Islamist pronouncements. She points out that women typically appeal to this ethical strand when they claim that Islam is nonsexist in its original writings (Ahmed, 1992).

These women claimed that Muslim women's pursuit of knowledge is not recognized in the wider society. On the contrary, they felt that "other people portray Muslim women as not technical." They felt that they face the prejudice that "we will be totally oblivious to anything outside our home," or even that "Muslim women won't be able to read or write." Seeking knowledge is, however, an activity that has limits; for some women, it will be done only with the husband's approval, or only by attendance at the mosque or the local Muslim Center. Thus, one Muslim woman was unable to continue coming to our Women and Science course because of family pressure on her "not to bother with that."

Like other women, they are concerned about environmental issues. In the group interview, they argued that Islam preaches, "mankind is a trustee of the earth . . . we are allowed to use the animal kingdom, [but] we have responsibility [even] for every twig we break . . . [if animals are to be used in science] then it must be in the best way." Better, some feel, to use human beings if products must be tested. Pointing out that much testing is done for "financial gain," one woman notes the way that science fails to give credit to other knowledges: "often they already have the numbers, the research, if they were looking around for it, in the regions of the world, they would have it there, the numbers and whatever they want." Here we see a clear echo of much feminist writing on science.

They are critical, too, of the way that "environmental problems today are to do with overproduction, with overusing . . . as Muslims, we're not supposed to live in too much luxury, to use only things we need . . . it's the illnesses of society that are causing the environmental problems."

Like many other women, they are concerned about the future directions of biotechnology and scientific interventions in human reproduction. They object to scientists "playing god," and to the possibility that people might be having "designer babies," aborting fetuses if they did not have (say) "blue eyes or blond hair." Part of their concern focused specifically on religious ethics. As one woman explains:

*a lot of science today is science for science's sake, they don't discuss the moral, the outcomes of the scientific research. They just do research, and find things out, but they don't consider the moral side . . . when they use the eggs of aborted fetuses . . . if someone is dead, even a fetus, that is still a soul, even a human body, once it's put into the ground, nothing should be taken away from it. We believe that the body still has some part of you, you're not away from it until the [day of judgement]—a fetus is still a soul, still a human being.*

Given that belief, they do not in general support abortion, unless the woman's life is in danger. They certainly do not support it in order to terminate a pregnancy on the grounds of genetic disease; the prevailing

belief here is that if Allah intended to give you a disabled child, he would do so. Abortion on the grounds of genetic disease, suggests one, "also devalues disabled people in society . . . if it's only two years [of life], but two years can be very valuable to people around that person . . . even if it's really disabled." Rather, they see having a disabled child as a test, "to see how we take care of this child."

Genetics, they felt, is too narrow a focus:

*. . . if you look at the environment, and [mal]formed babies and that, we run the risk of aborting everyone that's wrong, we don't see what's going wrong in society, where's it coming from? It's an easy option for industrialists—where do you draw the line? What is an abortionable deformity and what is not?*

In vitro fertilization (IVF), too, is not felt to be acceptable, as it meant not accepting what Allah had determined for you. Islam insists that a child is born of woman's body, which they interpreted as meaning all the stages of pregnancy from conception to birth. So, in IVF, "you're fooling the woman, because if it didn't get fertilised in her body, then she hasn't had a natural pregnancy anyway, so she's fooling herself that it's all my child." They are equally concerned about preimplantation diagnosis, in which one cell might be removed from an early embryo at the eight-cell stage. Wouldn't this, they wonder, involve a risk that scientists themselves are creating deformity in the embryo? Concern for the soul comes in here, in the context of a clearly developmental (epigenetic) account of the embryo that flies in the face of modern insistence on genetics: ". . . the soul is the soul . . . when you're talking about a person, if you're going to tamper with it at the beginning, you've changed it as a person; it doesn't matter about the soul [at this stage of embryonic development], you've changed it as a person."

The relationship between the soul and the body in Islam is, they feel, problematic for the practice of dissection. The body still has some part of the soul, even after death, so that "when a human dies, they're not really dead, but they can feel everything, when you cut them up they can still feel it." Another stressed that, "in the Quran it says, the body can feel everything . . . we haven't got the right to take anything, a heart, the liver, kidneys, anything."

Nothing can be taken from the body, for example, the placenta must be buried after childbirth. "You are not supposed to tamper with the body after death, even to the extent of cutting nails . . . your body might disintegrate, but the soul is still there, it will be given a form, even on the [day of judgement]." Indeed, the different parts of the body can speak in this final incarnation; thus, "God will make us speak out . . . if I deny [stealing] but your hands will speak out and say, but you stole with me, you made me do it."

Dissecting, one woman feels, is a "very intruding way of learning things . . . we've forgotten things, maybe there would be another way of getting around it." But, asks another, how would we know the Quran is right [about embryonic development, for instance] "without looking inside the body?"

Descriptions of conception and embryonic development in the Quran are now being reclaimed by some Islamic scholars as showing how well developed is the understanding of science quoted there. Certainly, Quranic stories of conception and human reproductive physiology are relatively egalitarian (Ahmed, 1992), unlike many accounts in western biomedical history (Tuana, 1989). Some scholars now believe that the Quran accurately described the way in which human embryos develop long before these processes were described by science. For example, in outlining the ways in which he believes the Quran to have preempted modern scientific accounts, Barr (1986) quotes Sura 23/12–14 of the Quran, which states that:

*We created man from the quintessence of mud. Thereafter we cause him to remain as a drop of fluid (Nutfa) in a firm lodging (the womb). Thereafter We fashioned the Nutfa into something that clings (Alaka), which We fashioned into a chewed lump (Modgha). The chewed lump is fashioned into bones which are then covered with flesh. Then we nurse him unto another act of creation.*

Scholars are divided, however, over the issue of when the soul enters the developing fetus (as, indeed, they have been in the history of Christianity) (Barr, 1986).

The spiritual soul was also a concern in our discussion about gene mapping; none of the women could really see the point, not least because it would not be possible "to map the soul." But biotechnology poses other hazards for Muslim women: the creation of transgenic organisms in particular is problematic because it means risking putting into one's body something that is forbidden. Anything with genes (or an organ) from a pig, would, for instance, be proscribed because eating pigs is forbidden on grounds of their being diseased animals.

Using human genes also posed potential religious problems, as the following extract from the group's discussion indicates:

*"I've heard that they are actually injecting human genes that produce milk, into sheep, so that they produce human milk. . . ."*

*". . . but then if you ate that meat, you'd be cannibal . . . which is obviously not allowed in Islam."*

*"but would it really be like human milk—how could it be?"*

*"I've heard of them putting pig's genes into other meat, as well, which causes a problem to us because we're not allowed to eat pigs, so if a little bit of the meat is, like, half pig."*

*"it's like, we're not allowed alcohol, but if you had a pot of milk that had just a drop of alcohol in it, we would not drink that milk."*

*"if you had [that] pot of milk, if the slightest impurity went in, you still know it's dirty."*

*"in our role as custodians of the earth, will we know what will happen once we start changing the plans, once we start making hybrids that nobody knows what the reaction will be, because Allah has made everything in a gentle balance, the whole of earth, of nature, is in balance."*

*"how can they ever make them safe? What if those sheep come out and mix with other sheep? Which sheep is a bit human and which one is not?"*

The question of possible risks to health if genetically manipulated organisms "come out" is certainly an issue that has worried many people. This conversation also illustrates a powerful belief in genes as essence; thus, a gene from a pig will retain some "pigness" about it, thus contaminating the meat of the animal into which it was put. This is not, of course, how geneticists would usually see it—to them, DNA is simply DNA. This belief is not, in our experience of teaching about "the new genetics" to adult communities, specific to Muslims, but is widespread. It is also not addressed by the scientific community.

Several other features strike us about this discussion. The first is that the picture painted by these women accords with feminist accounts of science. Sandra Harding (1992) has noted how the science that comes from non-European cultural histories often mirrors feminist accounts in its rejection of narrow reductionism. Afzal-ur-Rahman (1981) suggests, for example, that Islamic scientific approaches to biology are more holistic and provide more of an "organic unity" than western approaches.

The Muslim women here are similarly antagonistic to the reductionism of science. They clearly supported an epigenetic and transformational view of embryonic development rather than a limited genetic account (just as feminist biologists have done: see Hubbard, 1990; Birke, 1986; Fausto-Sterling, 1992). For these reasons, they are hostile to research into human gene mapping, which they see as explaining nothing important about humanity. Rather than accepting the view of some enthusiasts for the Human Genome Project that gene mapping will help us to understand or cure all social ills, they believe that children are "born pure" and that social problems have social causes (see also the critique of the gene mapping project in Hubbard and Wald, 1993).

Second, this material underlines the problems of assuming that "the public" is deficient in knowledge of science. These women are quite well informed about a number of contentious issues. However, they inevitably interpret these issues and information *in relation to* their worldview, their religious beliefs. Thus, their views about transgenics, embryo research, and dissection are all a complex product of learning some (conventional western) science and learning about their religion. So, while there were differences between them, these women tended to produce a consensus around how to interpret issues in relation to Islam.

Taking women (or, indeed, any adult) seriously in science education must mean taking into account these divergent, and sometimes contradictory, views. The position(s) these women take in relation to (say) scientific knowledge about the body has considerable implications for both health care and for the teaching of science at all levels.

Third, the local Muslim community in Coventry represents in itself a community that creates and labels knowledge. Not being Islamic scholars ourselves, we cannot tell whether all the claims made about, say, the environment are indeed supported by the Quran. But what matters here is the standpoint of these women. For them, Islam provides an interpretive intellectual, social, and ethical framework in which they must negotiate meaning. It is that framework within which they see phenomena or events as scientific or as important in their lives. Science must be studied, given the Quranic command to people to study well, but some parts of it must be assimilated to Islamic beliefs. Some elements, such as the theory of evolution, must be rejected.

Nonetheless, it is in relation to the role of Islam in the history of science

that we can see ways of improving science education—making it, as Sandra Harding (1994a) has emphasized, more multicultural. For example, chemistry is rarely (almost never) taught with much sense of history, other than reference to Mendeleyev as discoverer of the periodicity of elements. Yet Islamic science has been of great importance in the development of much of what we now know about the behavior of acids and alkalis (another Arabic word), and in the development of apparatus for techniques such as distillation (al-Hassan and Hill, 1986). These histories should be woven into the teaching of science if it is to be more fully representative of all our lives.

* * * * *

These case studies illustrate specific themes that emerged from the different social groups. Such differences between groups and institutions, we believe, have important implications for adult education practice and public understanding of science. The differences could simply have resulted from the sharing of ideas about the research before we arrived, to create some consensus and common discourse. In itself, this illustrates an aspect of experiential learning that is not adequately addressed in the literature on adult education—that the group itself is an important part of learning. We believe, however, that these differences reflect more than pre-interview discussions.

Perceptions of science are socially constructed. What is interesting to us as adult educators is how the various adult education contexts and institutions may have affected the women's views of science. This may happen either directly, through a women-only science course, for example, or indirectly, through involvement in a non-science-based course or an institutional culture influencing the women's view of knowledge and themselves as knowers. It may be that experience of some adult education counteracts dominant ideologies of knowledge (those that undermine women's confidence, collective identities, and claims to knowledge) in ways that may be very relevant to science education for women. Indeed, we believe that the views expressed by the women in our research study reflect not only perceptions of science in western culture, but also the ways in which particular standpoints and ways of knowing can emerge out of engagement in particular communities.

It seems unlikely that women's perceptions of science and scientific knowledge can be easily separated from their general perceptions of knowledge or of themselves as "knowers." It is probable, therefore, that their experience of adult education—even if science is not a part of it—could have a significant bearing on these perceptions. Adult education's sharing of ideas and experiences may encourage individual and collective thoughtfulness and critical reflection. This, coupled with the importance of the group and its dynamics in learning, may be more significant than the specific content or emphasis of any particular course, community group, or institution. The role of communities and groups in creating knowledge(s) has been largely neglected in debates about the public understanding of science that typically construct "the public" in terms of deficit or simple lack of scientific knowledge (Wynne, 1992).

Some of these points apply to anyone entering adult education. But there are ways in which gendered experience, of being women, will shape the dialogue between perceptions of knowledge and sharing of experience. This relates to Adrienne Rich's belief that the first liberating lesson in any woman's education is that she is capable of intelligent thought. Our reading of the evidence suggests that most of the women we interviewed had learned this first lesson (they are, after all, women who have made a decision to seek education). Many do respect their own minds and ways of knowing. The important point is that although, for example, most of the women see themselves (realistically) as passive consumers of science, many do not construct themselves as such. On the contrary, they see themselves as active knowers—even if not in relation to what they regard as scientific knowledge or what others would regard as scientific.

Our findings contrast with the mainly college women who were interviewed by Belenky et al. (1986) in their study of women's ways of knowing. Many of these women had not yet learned to see themselves as capable of intelligent thought. It also contrasts with Kim Thomas's findings regarding young women's experiences of higher education. She found that exposure to higher education actually increases the underconfidence of many women instead of challenging it (Thomas, 1990). The ability of the women we interviewed to see themselves as active knowers could, then, be a testament to adult education for women. It could also reflect that the very act of making the decision to return to structured learning enables women to construct themselves and their futures. In the next chapter, we begin to analyze in some detail the themes that emerged from our discussions with women about science.

GETTING THROUGH A DAY IS LIKE A SCIENCE
FOR SOME WOMEN. —BELL

# *The Research in Context*

## Introduction

The above quotation, drawn from one of the case studies in the previous chapter, represents many layers of women's lives. Work with adults has traditionally emphasized the importance of building on prior experience, but how can such practice deal with the relationship of adults to the abstract knowledge that we call science? What prior experience could be incorporated? And whose? Our research asks, first, how *do* women relate to this knowledge we call (natural) science? What does the word "science" convey to them? Second, what can we learn from what they say about the public understanding of science? And third, how might these perceptions contribute to the development of a more adequate science education for women and other excluded groups?

In this and the following two chapters, we look in more detail at the interviews and the themes that emerged from them. A theme of this book is women's shared experience of discursive marginality: they have on the whole been excluded systematically from theory formation and from "naming the world" from their experience. This, too, is likely to shape women's perceptions of science in fundamental ways. In this chapter, we concentrate in the first part on tensions and recurrent themes in the women's conversation in an attempt to map out some of the processes by which they saw themselves as excluded or marginalized by science. In the next chapter, we focus on women's responses, on the ways that they might express "not knowing," and on the problematic ways in which that not knowing might be interpreted. Finally, in chapter 6, we consider the multifaceted ways in which women articulated their ideas about "nature" in the interviews—the nature, that is, that science purports to study and explain.

The processes by which women are often excluded from scientific knowledge are complex and women do not experience exclusion in the same way. Nor do we claim that these processes of exclusion are peculiar to women: indeed, many men, from non-white ethnic groups, for example, are also excluded. However, we found that the language used to express their exclusion often represented something about women (invoking science as a man's world, for example) or it related to women's continuing role as primary caregiver in the family. Thus, anxieties about future technologies were often expressed in terms of what might happen to their children; others referred to the involvement of science in domestic work and others still to the position of women in the labor market.

Given women's relationship to the family and to the labor market it may be that women's relationship to science is in some ways different. It may be, for example, that the development of less detached, less controlling, and more responsible ways of knowing and dealing with the world become possible—as some feminist authors have suggested (Hart, 1992; Rose, 1994).

## Common Ground, Common Sense

The story the women tell us is certainly one of contradiction and exclusion. At the same time it is one of commonality through difference and it serves as a reminder that women seldom slip easily into their roles

as women, whether they be black or white, middle class or working class, young or old, able-bodied or not. The women's voices quoted here suggest some common ground; they also make visible differences between women rooted in different social experiences. In this chapter, we explore some of the contradictions emerging from women's speech.

In the women's talk about science, some linking themes emerge. Perhaps the strongest is the impressions that the women give to us—in their body language, their pauses, their silences. "Science" seems to evoke powerful memories, for many, of school science—experienced as deeply alienating, or at best as irrelevant.[4] Some of these memories are simply painful recollections of experiencing school as an alienating place, where one was rarely accepted. But, insofar as part of the painful memory for some women had to do with the feeling that they could not "get the right answer," then the apparent certainties of school science teaching produce a special kind of discomfort. Indeed, one woman began to weep as she recollected what school science lessons were like; she explained that the very situation she was in—answering questions—combined with the specificities of what she had to recall (science) brought it all back to her.

Such classroom memories structure how, as adults, these women perceive different areas of science. Thus, Chris can say of biology that it is "less scientific because I understand it. I put chemistry and physics on a very high level and I can't even *think* of the kind of person who'd go in for physics. . . . Medicine is very scientific because it's intellectually demanding." Similarly, Isabel expresses the same separation of herself and what she knows from the real thing when she says "Once you know about something it's less scientific."

How these processes of gender and exclusion work out in memories of the childhood classroom is further illustrated by the following extracts from our interviews. Here, the women concerned were discussing what they remembered from school about physics:

*I did start off to do physics [but then dropped it] . . . there were teaching problems . . . we were trying to struggle along with a teacher [who] sat up there just teaching us from books. . . . it was very mathematical and at the time I was struggling with my maths and I just couldn't get to grips with it . . . and I was, you know, a total failure. (Denise)*

*[Interviewer]: What about physics?*

*[Answer]: Honestly, I just draw [a] blank. . . . the only thing I can*

*remember from physics is something about having to measure certain things via mechanics, I can't remember.*

*I: Why do you think you can't remember?*

*A: Probably because I wasn't comfortable with it, I blacked it out. . . . I was naughty at school and I don't think I was encouraged enough in classes . . . I remember there was a basic interest, but I distanced myself from the subjects . . . because I didn't think I was any good at them. If I got one thing wrong that was it. (Tania)*

These extracts show the women concerned as feeling that they were not "good at" the subject, that they did not have what it takes to do physics and/or math. The contrast between them is also noteworthy, in that while the first implies a sense of *being* a total failure, the second conveys a sense that she deliberately distanced herself, a process in which she redefines academic "failure" as something in which she actively participated. This important distinction is one to which we will return.

Racism also structures experiences and recall of schooldays. For Gita, an Asian woman, the problems had partly to do with institutionalized racism and partly to do with language. She had always felt like an outsider at school, neither black nor white, not quite Asian but having Asian roots, always having to deny her background, even to the extent of "never [leaving] my clothes around at home" in case they smelled of curry.

Language, however, was a significant problem both for Gita and for her mother who had had an unnecessary hysterectomy because of language difficulties. Gita was subjected to IQ tests that were culturally biased:

*. . . they were asking me things about farming and vocabulary which I never come across . . . because to me that was [a] different language. . . . I mean even mixing with [white kids] my English was totally different to theirs, mine was an inner city sort of broken English . . . where I've been brought up with two languages. . . . there's a blockage, I wouldn't say I'm Asian, I was white, I was treated like white with the [other] girls and that I was fighting for my identity so that's my way of surviving.*

The language problem hindered her science studies, too:

*. . . because they were using words and the kids were used to it, and also they had the books and the knowledge they developed quicker, and I didn't, so that was the blockage.*

Gita sees her primary problem as that of language, while for many others it is the inherent difficulty of science. This theme is summed up eloquently in the quotation at the beginning of this chapter. Primarily, this expresses women's daily existence as often very difficult, a perpetual struggle. "Science" epitomizes difficulty. It also symbolizes the sheer drudgery of everyday work: for several women, science means boredom, a plodding approach to solving problems.

Some women explicitly express being excluded from science and scientific knowledge, demonstrating a clear awareness of powerful institutional, cultural, and educational processes that continue to exclude women. The young black woman Bell explains: "I feel so excluded from it. Science at school is made so impersonal and girls are told they don't have the capacity for it. That's a load of rubbish. And people don't know black people's contribution to science. As a black woman I don't want my children growing up to think black people have played no role in shaping today's world." In voicing a conscious awareness of exclusion she is also voicing resistance.

More commonly, however, exclusion emerges through recurrent tensions. While the women we interviewed see science in many contradictory ways, two particular tensions stand out. First, when women label knowledge as scientific it usually means something they do not understand: what they understand, by contrast, is likely to be labelled common sense. So, what they know becomes owned knowledge and more difficult knowledge is owned by other people. The second tension can be summed up in the aphorism: "Science is in everything, but it has nothing to do with me." This expresses the feeling that, although science is all around us, it does not feel as though it is entering their lives. Taken together, these themes serve as a stark reminder of the extent of women's exclusion from scientific knowledge. We explore these tensions below.

The interviewees bring out an opposition between common sense as "things we know" and science as "things we do not know." This tension comes out most clearly in discussing how they would judge topics and subjects in terms of how scientific they are. We asked the women to say how they would deal with a standard survey question that asked for a rank ordering (from one to five) of subjects like history, physics, chemistry, astronomy, astrology, psychology, biology in terms of how scientific they were. One woman's response sums up a commonly held view: "Well, I suppose what I would put as [number] one would seem the hardest to me. In fact that is the scientific one automatically to me and the one I understand is not the scientific one. History. I think I don't know anything scientific, so I would actually use that, what I think I am capable of understanding, as a scale for measuring it" (Isabel). Similarly, in the pilot study of "Inside Science" (referred to in our case study, above), asked to rank topics on the course in terms of "how scientific" they were, one woman says of medicine: "I don't know enough about it. . .so it must be science" (Birke, 1991).

It is in "common sense" that the women identify what they already know as not really science (for example, nutrition). Through this opposition, they locate themselves outside what they label as science. Veronica, for example, studying literature, speaks of people's "instinct . . . compared with science," on the grounds that "with science you have to have the knowledge." Formal knowledge of science is thus once again contrasted with common sense. As Isabel says, "once you know about something it's less scientific."

An overlapping issue is how women perceive the relevance of science; they consider those areas of science more obviously linked to the "common sense" of everyday lives to be the interesting ones. Thus, Veronica recalls doing experiments with reflection in physics at school and speaks of how this seemed more relevant than chemistry, "because it was there and you could see and chemistry wasn't something you could see. . . . It wasn't part of everyday life, it was something set aside that you had to learn about." Kirsty, in an introductory health course, feels that physics is irrelevant—a reason for her failure to remember much about it: "unless it was related to our life . . . now I can't remember laws and that I wanted to but my brain wasn't ready for it."

What we suspect is happening here is only partly to do with the specific subject matter of chemistry or physics. The feeling that something was "set aside" in the women's recollections may well reflect the way the subject was taught. Thus, the first of these women, unlike many others, felt that physics *was* relevant to everyday life, whereas chemistry was set aside. Clearly, what she was taught about light and reflection had meaning to her, while learning about the behavior of particular elements or acids and bases did not.

Making use of science and technology is seen to differ from understanding their basic principles; several women distinguish, for example, between using science or technology when they are using the products of previous research (such as domestic appliances or medical treatments) and understanding the principles of scientific knowledge. The former is

something to which they can relate; the latter is something from which they feel out of touch. Thus, sixty-six-year-old Lynn says she thinks she takes a lot of science for granted: "But I don't think I understand a lot of what I use. And I do like to use the latest models of things. I like it there but I don't think I understand much about it really."

The tension between science as everywhere, yet "nothing to do with me," comes out in an exploration of the various contexts in which women would describe something as scientific. Most women resist the "scientific/nonscientific" distinction that we made in trying to explore this theme on the grounds that "science is everywhere." For example, Sandra, a member of a women's health group, feels that "In a strange way science is around us all the time—in how we live, in hormones, in vegetables, pollution in water."

Similarly refusing to separate science from nonscience, Barbara believes that "whatever context you're in science is there. You can't put it in boxes. You can't have science versus nonscience. That's just human instinct, nonscience." Anjana, an Asian woman who came from Uganda, expresses the view that "We're surrounded by science; everything in ourselves, our way of life, psychology and computers, is related to science . . . everything to me is science." And Catherine, from the women and science group, insists that "Science is all around us. . . . Can you say a cup of tea is scientific because it takes a formula to make it?"

We take this to have two meanings. First, a scientific explanation can probably be found for just about anything—why the table stands up, or the appearance of colors in a painting, for example. Second, science's effects—say, pollution as well as modern technology—are integral to our way of life.

Yet at the same time, interviewees often express the "contradictory" view that science has little or nothing to do with their own lives. Gaynor, living in a rural area and a member of a health group, labels many of the things she does herself as scientific, such as cooking, or looking after her animals; yet at the same time, she feels that "Science just doesn't interest me. I'd rather leave it to someone else." Another woman's sense of alienation from science takes a different form. Claiming a fair amount of scientific knowledge in her everyday life, Barbara implies that this derives at least in part from her experience of growing up in Jamaica and of the black community there. She expresses the view that: "We are all part of science. We all take part in discoveries; we help one another find out why a tree is dying or what's wrong with us without going to an expert. And at home we figure out how to mend things. But our views aren't respected.

Science is something everyone should know. It shouldn't be segregated. We'd be more confident if it wasn't left to the white coats."

Science affects everything in our lives: "We can't get away from it," asserts Edna, including many daily activities in the category—vacuuming, using electricity (except cooking which is "sheer pleasure"). Yet for her scientific knowledge is: "all the things I don't know the answers to, things that don't seem to have answers—and I don't want to know." For her, moreover, "Scientists and food have become the new religion. We've replaced faith in God with an obsession with health and living long."

Inclusive versus exclusive definitions of "science" and "scientific" jostle with one another in interviewees' accounts. What the women include in the category of "science" depends on their beliefs and values. Some women who speak of science as "bad" then exclude from it what they value—natural history and nutrition, for example, are seen by them as outside the realm of science. Pleasurable activities cannot be science; thus, Edna goes on to explain that "people would argue that [cooking] is scientific, but as far as I am concerned it isn't, because it is a great pleasure." It is little wonder that Edna also felt that she couldn't "think how they can get the same pleasure out of say discovering an atom as you can out of reading [literature] which gives one's soul a pleasure."

Many women clearly expressed their feelings of exclusion from science. Thus, although Barbara acknowledges that she has some scientific knowledge from her daily life, she comments: "But I don't home in on it. It's the hidden part of you. I see talking about science as a man's world and you're not taken seriously if you talk about these things." The exclusion here recognizes the gendered construction of scientific knowledge that feminists have often pointed out (Keller, 1985). We pursue the theme of silence as an epistemological strategy in the next chapter. Sometimes silence is associated with fear, which may be vividly expressed in interviews: "With science it's a fear of not knowing, a deep inner fear of not being able to do it. I panic. With science I never questioned it because I was so afraid. It can still bring tears to my eyes. I'm still fearful of being put in a situation, even this one, of being given a puzzle to work out" (Denise).

## Science as (Un)certainty

Constrasting images of science as certainty and science as contingency also appear; "there aren't any right answers to anything" may coexist as a belief alongside a view of science as facts (right versus wrong). Recog-

nition of the changing, contingent nature of science makes no difference to the women's sense of exclusion because they have no part in any change. Negotiation of "the facts" is the prerogative of experts.

Research into how different social groups understand science has shown that people are often quite well aware of its contingencies and uncertainties (Wynne, 1992). Collins (1985) observes that the model of science as certain and of scientists as authoritative tends to be reproduced within normal science teaching. He suggests that in science research papers, "certainty increases because the details of the social process that went into the creation of certainty become invisible" (p. 160). The science paper leaves out much of the process of experiment. Science teaching and scholarly science papers are a fraud, he believes, at odds with actual practice.[5] And once those data find their way into textbooks they become absolute certainty, all contingency written out (Latour, 1987). This model of science allows lay citizens only two responses to science—either awe at its authority or rejection—"the uncomprehending anti-science reaction" (Collins, 1985, p. 161).

Interestingly, a small number of women referred explicitly to the central role of writing papers, rather than experiment, in the doing of science—just as Latour emphasizes. Thus, Rita felt that scientists were always trying to prove something, whether that is right or wrong, but, she stressed, what they produce is the scientific paper, in which

> *sometimes they contradict each other, what one scientist finds and they come to different conclusions. . . . I think it's normally, ummm, it's like an opinion really, everyone has got their own opinions and everyone has got their own way of going about things. So you are bound to come to different conclusions in the end really. I think that's part that is just normal.*

Here, Rita is clearly expressing the uncertainties and contingencies of doing science.

The notion of science as a social and cultural (human) activity is, however, missing from a minority of accounts that tend to conflate science with the world or nature. Veronica expresses this in terms of an opposition between "science" and "culture": "Science is finding out a fixed thing. English and history, say, can be picked up but science has to be taught." For her "culture [by which she means literature, history, etc.] relates to feelings. Science is rocks that have been found, the atmosphere . . . astrology [*sic*] and physics are out there."

For some, science included all kinds of intellectual activities. What mattered to Elinor was whether it involved lots of studying: commenting on the list of topics (history, and so on), she said,

*You see, I'd say all of them are science, even history—[because of] carbon dating—and astrology because it involves a lot study and it's gone on for such a long time which gives it credibility.*

By contrast, Monica, who felt that "we couldn't do without science," assumed a very different, and more exclusive, meaning: "Science has created a lot of monsters and without science we won't be able to undo the monsters." For her, science is about controlling nature, yet "Nature is as Nature is." This echoes what some radical science critics have maintained: that nature cannot be controlled—only "artificial nature," itself a creation of science: "What modern science may be capable of achieving is correctional hypotheses for earlier erroneous ones. . . . A great deal of science, then, is circular science" (Nandy, 1988).

Edna, despite her desire to claim pleasurable activities as outside science, claimed to admire scientists "more than all others." For her, experiments are "using knowledge to prove a point." Moreover, "people are overpowered by it, baffled by it, take it on trust and later its theories are disproved." For her, then, science is both certainty ("proving a point") and contingency. Perhaps it is little wonder that she thinks that reading science must be boring.

The contradictions between the certainties and the contingencies of science are brought out by a younger woman, Chris, who is concerned about the ethics of using animals in research: "We look to them [scientists] to give us rock hard answers, if you get a scientific answer you expect it to be the truth." Yet she goes on to point out that she feels that testing drugs on animals is inappropriate because the answers are not always correct. When asked about the contradiction, she suggests that perhaps she has not "got my facts right . . . I think maybe I've been brought up to believe a scientific report. Maybe because I've got this idea that there is lots of test tubes and bunsen burners and that's what they're doing and then maybe I can believe that and what they are testing is right. But where I was actually looking at the side where it wasn't bunsen burners and test tubes any more, it's animals, that's when my opinion changes."

Another woman, in the Women and Science course, notes the contradictions in her own thoughts about science. Asked what she understood by scientific knowledge (a question most scientists would find difficult to

answer)—she replies, first, "It's hard facts. It's all definite, whereas, say, common sense and logic have shades and compromises. There's no in-between in science" (Carla). But when discussing how scientific knowledge grows through experimentation she shifts her point of view:

*Interviewer: So, if there's disagreement it's because they haven't done enough to get the right answer?*

*C: No, it depends on the interpretation of who is doing it. I could see this, you the opposite. But you've still got the facts. But that's contradictory, isn't it? But you've still got the facts, haven't you? I mean, what's there is there. They're still black or white.*

*Interviewer: So science is about "the facts" but it can only interpret them?*

*C: It should only be one interpretation, though, shouldn't it? Yet everyone could have a slightly different interpretation. Their logic or common sense comes into it. Yet I didn't say that originally, did I? I said science to me is hard facts and logic and common sense are shades. . . .*

The stereotype of scientists as "other," unlike ordinary people, is also common. This stereotype is used even by the women in the Women and Science course. The women tutors are seen by Catherine, for example, as "different, quaint, really"; scientists in general she sees as removed from reality, separate, even "mad" and "very intelligent." (We should note here that it was not clear from this interview whether she felt that the tutors were "not like ordinary people" because they were scientists, or because they dealt with a feminist curriculum!)

Sometimes gender is mentioned (scientists, unsurprisingly, are men); race and class come up less often. We would have expected, given the research focus on *women's* perceptions, that gender would be foregrounded. But when speaking about scientists there is a tendency to describe them in personal, psychological terms rather than social ones. Common descriptive phrases in questionnaires and interviews include "brainy," "boring," "unintuitive," "enthusiastic," "curious," "out of touch," with only a minority using categorizations like "white" or "middle class."

Despite a fairly widespread awareness that science is a social and political process that is influenced by business and government interests, perception of the quest for scientific knowledge as an essentially individual endeavor emerges from the interviews. For example, the "common sense" referred to by the interviewee above is attributed to scientists as unique individuals, rather than as members of a narrowly based, privileged social group. This notion of scientific knowledge as something acquired by the few, very intelligent among us could be graphically expressed: thus, asks Chris, "How can they *hold* all that knowledge in their heads?"

In Chris's question, the individualized expert and his knowledge stand opposed to the more obviously collective practices of knowledge creation with which she is engaged as part of an explicitly feminist teaching program at Hillcroft College. There are two points of contrast here. The first is that, in relation to scientific knowledge itself, women typically portray themselves as passive consumers/receivers of knowledge, who might perhaps acquire odd bits of scientific information (reminiscent of the "passive knowing" that Belenky et al. [1986] described). The second is to contrast the expert and his possession of elite knowledge with knowledge that women (sometimes explicitly) feel they "own" and which "I'd rather have than all that science stuff" (Chris). Rita, for example, says, "I can only grasp an idea if I can put myself into it."

Referring to the "science is everywhere versus nothing to do with me" polarity, Chris goes on to say:

*I don't feel it personally . . . I know it but don't feel it . . . that science is everywhere. I pigeonhole it and see it as separate and that's how I divide it from common sense and women's intuition.*

Chris "can't even *see* a woman doing science . . . except on the caring side . . . can't *see* a woman with test tubes . . . on the intellectual side . . . I can't even *think* of the kind of person who'd go in for physics" (emphasis in her speech). For Monica, "Biology is easier. With chemistry you need more understanding and with physics it's pretty much all understanding. The more difficult, abstract, awe-inspiring, the more scientific. Anyone who can get their mind round that is very impressive to me. Boring in a sense . . . I don't really want to be able to do it myself."

In these accounts, what scientists do is perceived as boring, tedious, mathematical. While those with some connection with science tend to see what scientists do in terms of "the scientific method" (testing hypotheses, for example), others tend to see experiments as "luck, a matter of trial and error"—though this does not appear to affect their trust or lack of trust in scientists.

## Self and Science: Nonoverlapping Sets?

These various tensions powerfully reproduce and maintain women's exclusion from science. If science is all around a person, for example, then where is the self? If science is an individual endeavor, understood only by those eccentric but clever persons known as scientists, where does this leave women's knowledge or their "common sense"? Women's exclusion from science is not only sheer numbers who drop out of science education: it is more profound than that. It is fundamentally about the status and power of some knowledge claims over others.

The perception of science as the preserve of specific kinds of individuals also has implications for wider concerns over public understanding of science. If science is seen as individualized, then those perceptions must be strongly influenced by the stereotypes of scientists pervading our culture, as we noted in chapter 2. The overwhelming negativity of these stereotypes must influence how people see scientists; it must also influence perceptions of science itself, as a social practice. It is not only the "facts of science" from which many of these women distance themselves: it is also the arcane images of those eccentric scientists.

Some of the themes we have identified are not unexpected; it is widely known, for example, that women feel excluded from science. But what we want to emphasize here is that these themes are rooted in oppositions—between science and "common sense," between owned and alienated knowledge, between science as everywhere but "nothing to do with me."

In their account of women's different "ways of knowing," Belenky et al. (1986) distinguish several approaches to the acquisition of knowledge, ranging from the silent knower through the knower who actively constructs knowledge. Several of these are identifiable in the interviews, sometimes in the same person. The point here is that although many of these women have no difficulty in seeing themselves as active knowers and constructors of meaning, this perception seldom extends to the realm of the "scientific." Nonetheless, it would be a mistake to see the women we interviewed as necessarily or merely *passive* in their relationship to scientific knowledge. Science is for some women the antithesis of "real knowledge" that they can intuit or arrive at through their own processes of analysis. This is more meaningful than the kind of knowledge that science represents to them (a point explored in the next chapter). Often, this "owned knowledge" is related explicitly to their experiences in a racialized and class-conscious society. How it does so, however, can be different.

**Owning Our Own: Race, Class, and Knowledge**

Acknowledging "other" ways of knowing is a central message and focus of Wendy Luttrell's research (1989) on the relationship between working-class women and knowledge. Her work is particularly useful here for three reasons. First, it focuses explicitly on the troublesome and paradoxical situation many women face in pursuit of adult education. Second, her study shows how complex gender, "racial," and class relations of power shape women's perceptions of knowledge. Third, it validates what Adrienne Rich has said about women's education: the most important thing that a woman can learn—what she needs to know above everything else—is that she is capable of intelligent thought. This is, she believes, the first lesson in any liberating education for women (Rich, 1979, p. 240).

All of the women Luttrell interviewed distinguished between "common sense" and "school-wise" intelligence, that is, between knowledge produced through experience and knowledge produced in textbooks by experts. As Karen Brodkin Sacks (1993) has observed, working-class women's community cultures have "consistently reinvented a range of 'anti-bourgeois' structures and values to shape their communities and guide their lives" (p. 17). Appealing to "common sense" thus contributes to an oppositional consciousness, rooted in working-class experience.

Luttrell points out that working-class women share similar ideas about their commonsense abilities to care for others; they regard common sense as a way of judging truth on the basis of what trusted people have seen or experienced and know to be true. The claim to have commonsense knowledge, she suggests, recognizes and validates working-class solutions to problems despite the power of scientific knowledge: for example, relying on friends who know the ropes, and seeking advice from trusted people not because they are professional experts but because they share the same problems.

How to share and develop the collective knowledge that results from caring for others is a problem when such experiential knowledge is dismissed as purely subjective. Yet knowledge born of practice, as in midwifery and nutrition, is often more securely founded than the proposals from an often "arbitrary science" (Rose, 1994). In our own research many women dismissed what in other contexts would be labelled scientific knowledge—for example, knowledge of nutrition or cooking are regarded as "just common sense" or, even, "just sheer pleasure."

Appealing to "common sense" in these ways brings to mind what has

been called "really useful knowledge" in adult educational practice. This was a feature of informal, self-help working-class education, particularly in nineteenth-century Britain. In this tradition, self-education by working-class people *for* working-class people formed part of an oppositional movement with an alternative vision of society. Education wasn't separated from life in local neighborhoods, at work and at home; knowledge was seen as lying in everyday life as much as or more than in books (see Johnson, 1988, on this tradition in Britain, and Montgomery, 1994, on the place of science in education for democracy in U.S. history).

However, in Luttrell's study there were also important differences between women. She suggests, for example, that although both black and white women claim commonsense knowledge, they are distanced from their intellectual capacities in different ways. But for both, the ideology of "intelligence" acts as a filter through which they deny everyday experience and knowledge.

When asked about people they know who are intelligent, white working-class women in Luttrell's study referred exclusively to men. That is, they see some aspects of common sense as real intelligence but only those ways of knowing associated with men's skilled, manual work and abilities. They construe these in opposition to middle-class professional people (men and women) who, in the words of one interviewee, "don't know how to do the simplest everyday things or cope with everyday problems—that takes *real* intelligence; it takes common sense."

However, the white women never refer to women's manual jobs or abilities to work with their hands, as requiring real intelligence. And while they equate men's self-taught activities such as playing a musical instrument, with intelligence, they ignore the range of their own self-taught activities such as helping children with their homework. They similarly dismissed their own activities in the family or community as trivial. These they saw as acquired naturally or intuitively, unlike the men's craft knowledge, which is more obviously acquired through public, collective experience.

The black women Luttrell interviewed also locate their commonsense knowledge in caretaking and domestic skills performed for others. Like the white women they describe their common sense as "intuitive" and stemming from feelings; similarly, they focus on the common sense required to raise children. Luttrell comments:

*The [black and white] women's classification of their knowledge as "affective," not "cognitive"; as "intuitive," not "learned"; or as "feelings," not "thoughts" all reflect an acceptance of dominant conceptions of knowledge and ultimately diminish women's power. (p. 40)*

The learning process involved in acquiring commonsense knowledge—based on caring and relational aspects of the women's lives—remains invisible; this "intuitive" knowledge, suggests Luttrell, is individualized and personal, not collective or public. It is associated with feelings and intuition as opposed to thinking and learning and is experienced as affective, not cognitive.

Indeed, Luttrell argues that because women are not allowed, ideologically, to be sources or agents of rational, legitimated knowledge they associate the knowledge they do claim with feelings and intuitions. Both classifications ("common sense" and "intuition") place women in less powerful positions relative to men (both black and white) and to white middle-class professionals (male and female) in particular. And they do so, she suggests,

*not simply because women are . . . seduced into believing in the ideological split between feelings and rationality . . . but because the real nature of women's knowledge and power is hidden from view and excluded from thought. (p. 40)*[6]

However, black and white women do not experience their exclusion in the same way: "Race" influences how the women claim knowledge, as well as how they experience exclusion. Black women, for example, *do* claim their own common sense or "motherwit" as real intelligence and perceive it as based on the ability to work hard and get the material things they and their children need, with or without a man's support. They also regard their ability to deal with racism as another form of real intelligence which they share with black men against the ignorance of whites.

This relates to Patricia Hill Collins's distinction between knowledge and wisdom, and the use of experience as dividing them. Knowledge without wisdom is adequate for the powerful, argues Collins, but wisdom is essential to the survival of the subordinate (1990, p. 208). Wisdom is what the black women in our study, too, asserted for themselves.

Luttrell suggests that it may be because black women's work as women is also manifestly the work of black survival that it is not as easily trivialized as that of white women. But another, connected reason, we think, is the tradition of sisterhood in black communities that crosses class lines. Black women's centrality in families and in various community

organizations—as "the ties that bind the black community together" (Dodson and Gilkes, in Collins, p. 141)—provides them with a framework for valuing their own concrete, experiential knowledge. This allows them to share this concrete knowledge of what it takes to be self-defined black women with younger, less experienced sisters (see Collins, p. 211).

Such a sense of family and community and of shared responsibility for children is clear in the following description of Cariba, a black women's organization in Coventry, by one of its members:

*I see it as a family group . . . and it has a lot to offer the community. . . . As a black women's organisation I think we can influence our young black people, our children . . . and help them become adults in their own lives. (Martha)*

Collins believes, too, that dialogue, connectedness, and emotional investment are part of an Afrocentric tradition for assessing knowledge claims. So black women, unlike white women, may again experience a convergence of the values of the black community and women's experiences that enables them to claim their knowledge (see Collins, chap. 10). In fact, she believes, black institutions and families encourage the expression of black female power, partly through support for an "Afrocentric feminist epistemology" (Collins, p. 217). Few "Eurocentric" institutions, including educational establishments, value this way of knowing.

In our research, several black women clearly laid claim to their own (Afrocentric) cultures in a racist culture. Discussing notions of alternative health and diet, some speak disparagingly and dismissively of the ignorance and gullibility of people taken in by this (expensive) western fad; at the same time they point out that much of it is ordinary back home. The following remarks are typical:

*That's one thing you western people are crazy about. OK. It's important to be careful about what you eat but you can go too far. . . . Alternative things are very scientific but they've existed a long time. Some people here just want to make money out of it; at home, it's poor people who use it and who eat brown bread. Here, it's expensive and for the middle class. (Brenda)*

*It's nothing new. To be told it's new is quite puzzling. But it's always been there. We're just harnessing it with designer labels. (Molly)*

Luttrell believes that black and white working-class women are not distanced from their knowledge in the same way: the daily reminder of a collective identity as working-class blacks, she suggests, mitigates the daily reminder of their individual identity as women. *Their* intuitions and particular claims to knowledge of relationships are part of a collective identity as black women.

Our research reveals widespread anxiety about science "going too far" and beyond its understanding and ignoring the emotions. Questionnaire returns are replete with criticisms of science and scientists for being too abstract, analytical, academic, and narrow-minded and for failing to integrate artistic, imaginative, literary, evaluative, social, philosophical, personal, spiritual modes of understanding. Half the returns refer to the isolated, separate nature of science and over a third refer specifically to its separation from other forms of knowledge and understanding; for example, "its distance from social science, philosophy means lack of understanding of applications of science or of the interrelatedness of life" (LC); "Sciences don't deal with ambiguity, don't deal with human meaning" (E); "linear, cold, analytical, isolating" (Sh). Scientists are "inflexible, closed, lacking in true wisdom" (Sh), "have tunnel vision, unprepared to accept responsibility for consequences" (LH), and so on (letters in parentheses refer to groups—see Appendix).

Yet we found little evidence of any simple turning away from science. A much more insistent voice is for a re-formed science. It is thus not only exclusion being expressed, it is also anger and disillusion at the kind of science we now have. Dissatisfaction with the abstract, controlling power of modern science is clear.

In the next chapter we pursue the theme of challenge and resistance to scientific accounts and explore how engagement with particular communities influences the women's discourse. In this context, we illustrate how it is in voicing resistance that differences between, for example, black and white women's accounts emerge. We therefore shift our focus from similarities to differences, in particular to how women position themselves in relation to science and scientific knowledge according to age, race, and ethnicity.

Some of the contradictions we explored in the previous chapter serve to underline the ways in which women can feel outside science and its authoritative voice, while simultaneously recognizing its importance. The gendered character of scientific knowledge means that women's location always begins from outside science. But there are differences between women and different degrees of being outsiders. It is difficult indeed for any woman to become "inside" the practices and authority of orthodox science. It is even more difficult if she is not white or middle class.

Beliefs among the women whom we interviewed that science is "nothing to do with me" result from the elitism of scientific knowledge, and hence the marginalization of all kinds of people. But they are *also* statements about women's experience of high school education that gave them clear messages about science being "for the boys." Again and again, we heard remembered tales of women's feelings as they were told that science was not for them, or witnessed boys in their class demand, and get, teachers' attention. For many women, recounting those experiences of the school science classroom evoked painful memories, in which "women's words trembled into the . . . space" (Lewis, 1993, p. 164). A particularly clear example of how one black woman saw science as being male invoked a powerful metaphor of exclusion:

*I suppose it's white men in white coats—you know, John Major* type glasses . . . and good clinical white buildings with fences outside not particularly high with notices "Do not enter—restricted area" and that sort thing, do you know what I mean? (Bell)*

Given the impression that women don't belong in science, saying that "it has nothing to do with me," is a profoundly gendered statement. The women we interviewed had been educated in the British school system or its colonial derivatives in Asia, Africa, or the West Indies. The elitist model this system has traditionally espoused (particularly in England and Wales, perhaps less so in Scotland) has been supported by specialization, so that girls must choose subjects early in their teenage years, at a time in their lives when gender conformity seems to matter.

Those messages—about elite forms of knowledge and who can, or should, have access to them—are learned powerfully and early. They serve to position all of us. Yet the women we interviewed also positioned themselves, so that they rejected some knowledge that might be called scientific, and some they accepted. Only some scientific knowledge was clearly "nothing to do with them." As we have seen, how that knowledge is labelled is important; the women were more likely to claim it as their own if they could allocate it to "common sense."

In this chapter, we explore some of the nuances of our conversations with women, in order to identify a broad theme that illustrates, we believe, the complex ways in which women come to position themselves in relation to scientific knowledge. We consider here how the women use either silence, or the "don't know" response. Certainly, some aspects of the conversation were likely to trigger such responses as silence—the very

*The (Conservative) British Prime Minister at the time of the interview

word "physics" often did, for example. This reminds us of the "silent knowing" that Mary Beth Belenky and her colleagues refer to, the silent, passive student in the classroom. But silence is more than simply not responding or not knowing. It might also represent resistance—either to something associated with "physics" and its imagery, or perhaps to a research process that required them to think about it at all.

## Knowing ignorance?

We use silence as a starting point from which to explore what might be called "knowing ignorance"—an active recognition that one doesn't know and *does not want to know*. In exploring the issue of women's silence in the classroom, Magda Gere Lewis (1993) speaks of the tension between silence as oppression and silence as revolt (p. 2). Silence, she points out, is too often seen as women's fault, a product of their deficiency; yet it can also be a form of resistance to specific discourses. In our interviews, silence as revolt seems to be associated with particular groups. Building a discourse of refusal[7] around science seems to be part of specific epistemological communities, to return to that term.

Whatever official pronouncements may say about public (lack of) understanding of science, our conversations with women illustrate the complexities of their understandings and responses to scientific knowledge. Silence is one possible (and not unexpected) response to being asked to talk about science. But what does it mean? Silence may be a response to the wording of the questions or to the research process of interviews. As feminist researchers have often commented, there is a real problem for feminists in approaching research with women; how can the interaction be anything but oppressive or exploitative—especially when the topic being discussed is one that disempowers many women? On the other hand, these were women who had agreed to being interviewed about science, though they did not know beforehand what we were going to ask. Silences in response to questions usually did not last long. We could always rephrase the question or prompt in other ways. But they did tend to occur in response to particular words, phrases, or ideas.

Responding with silence may reflect the nature of interviews as dialogue, in which both participants have expectations. Several women interrupted the interviews by asking if their answers were "what we wanted." Not responding immediately by being silent may be one way of expressing uncertainty or that expectations of the interview were not being met. While these are generally issues facing any interviewer, they may well be more salient in relation to certain topics.

For younger women, "science" often evoked associations with specific subjects such as physics or chemistry, which have sometimes been taught as separate subjects in school. It was these words (physics, for example) and images of school laboratories, which most consistently produced silences or uncertainties during the interviews. Questions often provoked short, sometimes staccato, responses. Yet, moments later, the same women would become much more articulate, talking at length for example about their feelings and reactions toward food irradiation or about the politics of funding in science.

The brief silences (and staccato responses[8]) shifted the interview. Refusing to be drawn out on particular topics that represented something largely irrelevant to their lives moved the interview onto safer ground. For nearly all the women, thinking about, for example, what physics might contribute to daily life was difficult; it just seemed to have little real connection. In that sense, silence is not only a reaction to particular ideas or words, but is also a rhetorical strategy.

The meaning of silence, of not talking about something, as a strategy was sharply underlined by Barbara, who emphasizes how she might avoid talking about scientific matters to others because of how she might be interpreted in relation to race and gender:

> *I see that part of, umm, talking about science and things as a non-white woman. Even though you know if you were to discuss it with a man, umm, or in a group it wouldn't be taken seriously as a part of your everyday conversation. . . .You are taken more seriously if you are talking about going to the hairdressers or going on holiday. . .but how bacteria, how cancer, you know all the research, that kind of thing, is really left to a higher . . . [she drifts off here].*

Later in the interview, she repeats this theme. Having told us about a television program she had seen about pesticides and crop production and how interesting she found the topic, she says that "when you see things like that you automatically home in on it—it's the part of you that's the hidden part." She hesitates to share this interest even with friends. With friends, she notes, conversations would return to the latest episodes of soaps on TV, even when she knew that her friends were also interested in the pesticide program. It is as though to have serious debate is to threaten particular identities.

For Barbara, silence is an epistemological strategy shaped by race and gender; as a black woman in Britain, silence is less likely to lead to difficulties than speaking out. For her, it is also a strategy rooted in colonial history. Barbara was one of several black women we interviewed who had not been born in Britain (she emigrated from Jamaica twenty years previously, at the age of twelve). What is clear from these women is the extent to which they were exposed to "British" education and values in their countries of birth—that is, to an educational system built on colonial values and deeply rooted in elitism and social class. Studying science often meant—as it would have done for Martha, growing up in Ghana—travelling to go to the local *white* boys' school to receive an education rooted in assumptions of class and the "superiority" of colonial education.

The science that these women were taught at school was inevitably mired in Eurocentric assumptions; they were, for example, more likely to be taught about the four seasons of the temperate zones than anything about the seasonal patterns (and hence crop growth) of the equatorial areas in which they lived. Monica, for example, told us of her experience of learning science through colonial education in Barbados—here, she is speaking of learning about winter:

> *[I imagine I learned some of it in science] but my earliest recollections would be through, ummm, you know, poetry most especially. . . . I was in Barbados and literally we had two seasons, which was the monsoon season . . . and the summer. More perpetually summer, and I mean I learnt about snow and things like that, and autumn—you know, winter, autumn in a foreign country, umm, so the concept of snow, I mean, you didn't have snow in Barbados—so, ummm, that's the way I would have learnt about it initially. . . . I suspect that . . . science classes would have devised to telling us about how snow molecules formed, you known, crystals. You know, that star shape, you visualize stars at night and that's how it looks.*

For Monica, as for many of the black women, the irrelevance of much of science has a great deal to do with imperializing and Eurocentric assumptions made in the educational process. If women in general feel that much of what they learn about science is irrelevant, how much more so must the woman who learns a science derived from study of "nature" on another continent? Silence, in this context, is a response to an educational system that denies the existence of one's own experience and knowledge.[9]

As we have seen in chapter 2, some of the debate about public understanding of science seems to portray nonscientists as deficient, as lacking the "right" knowledge.[10] If only the public knew more, they would be more supportive, the argument runs. Yet "ignorance" is itself complex; it could mean many things. We might say "I don't know" because we simply do not know the answer, or did not understand the question. We might say it in order to protect someone else, or because we feel it is not our business to know. The "don't know" response might also be attributed to a perceived lack of confidence in one's own ability, not having the right "kind of mind." Or, as in the case of some of our pilot interviews, "not knowing" may be attributed to having "missed something"—a particular class on a particular topic, for instance.

In relation to science, Michael (1992) distinguishes between ignorance of science-in-general, and that of science-in-particular. The example he uses comes from his study of public understanding involved in a survey of attitudes towards testing for radon. People were aware of the purpose of the radon survey; they were supportive, that is, of science in practice. But they expressed no interest in the science necessary to explain what was happening. In other words, they were "ignorant" of science-in-general, the principles behind it.

Our interviews provide examples of the different ways of professing "ignorance." The need to have the "right kind of mind" to understand physics (or science in general) is a powerful theme (and it echoes a theme brought out in Michael's interviews, in which some people felt they did not have the right kind of mind to understand the physics of radon). Not knowing can then be a perception of self as lacking some ability: physics, again, is typically the trigger for this kind of response. Thus, Kirsty tells us:

*Oh, physics, oh I desperately wanted to know about physics, but to my horror of horrors, and my poor lecturer at school ummm I can't really say what I learnt in physics . . . . Unless it was related to our life, you know, it came into our life now I can't remember laws and that I wanted to but my brain wasn't ready for it.*

Sandy, a student in the Women and Science course, similarly recounts her experience at school of feeling that she "wasn't up to it":

*I did do physics at school, but that was sort of way over my head, I couldn't get into it [because it was so abstract], I couldn't see the relevance of that affecting my life later.*

Perceived relevance is an important issue. Most women felt that they knew little or nothing about physics and chemistry, partly because they had seen it as irrelevant to their lives when they were at school. Those few women who had enjoyed physics, and who expressed some interest in it, reported understanding it as relevant. Veronica, for example, regrets that she did not do a degree in mathematical physics— which wasn't possible, she said, "in my era" (she was 49 at the time of interview in 1992).

The arcane worlds of physics and chemistry are, for most of the women however, completely apart. They see no connection with their lives. Interestingly, few of them mention the need for high levels of mathematical skills to do physical sciences (although several feel that they "couldn't do" physics). Rather, the primary concern is with irrelevance (even though, contradictorily, some women admitted to fascination with astronomy). "Not knowing" about physics was firmly linked to a widespread perception of some areas of science as being apart from everyday life.

The issue of relevance is important here because it reflects a widespread perception. Yet, observing schoolchildren closely in their subject choices has led Joan Solomon (1993) to reject the claim that irrelevance is a reason why young adolescents reject science. If she is right, then what is noteworthy here is how these women construct a story, on the theme of irrelevance. Perhaps, following Solomon's research, irrelevance was not the main reason for their choices; perhaps, as she suggests, it had more to do with issues of learning style. But irrelevance is clearly how the women perceive those choices now. And now is a time when many of them have daughters and sons who may end up making the same choices.

## Lurching in the night: facing up to technology

Even when physics does begin to enter everyday life (for example, in understanding how some domestic appliances work) it remains something separate from experience. We might *use* an appliance, but not want to know how it works; or, not knowing how it works might be a stated reason for not having the appliance. Thus, many women told us that they did not really want to know how a microwave oven works, and some felt they did not want one in their kitchen for fear of what radiation it might emit. Again, this echoes similar findings from other studies: even those who work with radiation (such as apprentices at a nuclear power station: Wynne, 1992) express their distance from understanding the principles.

Technology, as several feminist writers have observed, is strongly gendered (L. A. Keller, 1992). In the workplace, technological innovation has all too often meant loss of jobs, particularly for women, while men have acted through trade unions to protect their interests around the new technologies (see Cockburn, 1985). Technology in the home may be designed with the assumption that a woman will operate it, but also that she will not understand it. This point was put clearly by Barbara, who talks about the scientific aspects to using a washing machine (for example, the way the detergent interacts with fabrics), but then goes on to make a contrast between something being scientific and women's role:

*you know, it is scientific, but you don't assume [that], because it's just an extension of woman's role, you just wash the clothes you don't see the scientific . . . you just know it gets the dirt out*

Here, "ignorance" of how it works is expressed in explicitly gendered terms: you "don't see the scientific" if it is "just" women's role.

Most of the women we spoke to didn't know how a microwave oven works. Here, again, is a rhetorical blocking, a clear closure to the conversation. A few said something about cooking from the inside, or mentioned waves, but were clearly uncertain. Silences and hesitances abound. But several speak more strongly, emphasizing that they do not like the idea of a microwave. Here, the answers are much more confident and forthright. Thus, Edna says simply that "I wouldn't understand how a microwave works, I mean, I am not very interested in those things," while Martha emphasizes that she doesn't like the whole idea—"just a feeling about it," she stresses.

Gita, similarly, speaks of her suspicions about machines in general and microwave cooking in particular:

*I have not got a microwave and I have never had a washing machine. . . . I am not a very machine type of person . . . never been able to afford it . . . I wouldn't know how to work [it] because I haven't had the upbringing. But, err, a microwave, I would just . . . read the instructions at the front if it's a vegetable but I would not cook meats . . . I am suspicious of it, very suspicious of it and I like to do my meat really well . . . I don't feel it's safe enough, I don't think they know enough, it's a new invention, just come in, and you don't know, in fifty years time, the disadvantages of it.*

Her concerns here are not only about the technology itself and whether she could use it, but also about not trusting expert opinion, which might gloss over potential hazards.

The notion that microwave ovens might pose not fully understood

hazards was a frequent theme (and an important one, to which we will return). But Selma takes her suspicions of the technology itself even further, creating a kind of machine-monster image:

*I haven't a clue about [how they work], I've no knowledge whatsoever. In fact, they frighten me to death . . . they alarm, I feel sure that they threaten me . . . they have got that look about them, oh yes, if I had a microwave it would sit in the kitchen waiting to attack, it would lurch forward one night and engulf me.*

An imaginative flight of fancy perhaps, but the belief that somehow machines can threaten is a powerful one (and not entirely misplaced; the development of microwave cooking came from military application. How much of that knowledge formed part of Selma's explicit or implicit world view?). "Ignorance" here is clearly not defined simplistically as "not knowing": it is a willful rejection of both the knowledge and the associated machine.

## Mistrusting Experts?

Yet alongside the statements of "ignorance" and the silences, there is also much anger and refusal of the terms of the discussion. Sometimes silence was about resistance, about refusal to engage in conversation about certain topics (women's body language often indicates that silence is not just about failing to speak).

Another facet of refusal is shown by the way some of the women express mistrust of experts, as Gita does in the quotation above. Scientists, thinks Edna, expressing her mistrust (alongside her awareness of the power of particular discourses),

*tend to dominate ideas about what life is all about because they do . . . produce such peculiar, I mean, the sweeping statements that are made due to the fact that they have done a piece of research and . . . out of that research will be extrapolated a whole set of ideas which then become dominant like the ozone layer. . . . I think most people are overpowered by a lot of scientific. . . .*

Such attitudes toward expertise may, paradoxically, coexist with attitudes of trust, especially if there is some kind of regulatory body. Thus, Monica is prepared to be trustful with regard to food additives, on the grounds that "everything [E numbers] has to be registered and things are probably fairly harmless."* Being registered seems to mean to her that the additive had been exhaustively tested and was therefore safe. Yet, a few minutes later, she expresses mistrust of expert pronouncements on washing powders that are alleged to be phosphate-free.

Mistrust of experts is a common response. In Beck's (1992) account of the "risk society" that he claims is beginning to supersede industrialized society, he suggests that people are increasingly aware of the environmental risks to which they are exposed. Today, we are increasingly exposed to risks over which *none* of us can have any direct control—the holes in the ozone layer, for example. Among the consequences of this, suggests Beck, are a growing sense of powerlessness and a mistrust of expert pronouncements on potential risks.

Mistrust of experts is not, Beck argues forcefully, based (as many scientific experts would have it) on public ignorance. Rather, it is rooted in people's recognition that expert opinion, based almost entirely on technical, numerical data, completely ignores the social context in which those very risks are experienced as well as the expertise of others who might have an interest. Beck puts it eloquently, speaking of parents in Germany who have watched their children become sick as a result of exposure to airborne pollutants, but who are now forming citizens groups to fight back:

> *They no longer need to ponder the problems of their situation. What scientists call "latent side effects" are for them their "coughing children" who turn blue in foggy weather and gasp for air, with a rattle in their throat. On their side of the fence, "side effects" have* voices, faces, eyes *and* tears . . . *and yet they must soon learn that their own statements and experiences are worth nothing so long as they collide with the established scientific naivete. The farmers' cows can turn yellow next to the newly built chemical factory, but until that is "scientifically proven" it is not questioned. (Beck, p. 61; emphasis in original)*

Expert opinion all too often relies primarily on the need for a particular *kind* of numerical evidence: mere correlation between two events, for example, is not considered sufficient, and contexts are ignored. Neither is particularly likely to appeal to the parents of a child dying of leukemia, having grown up near a nuclear processing plant. Experts claim that there

---

*E numbers are a system of classifying chemical food additives, used in the European Union.

is "insufficient" evidence to link clusters of childhood leukemia cases to nearby nuclear plants: mere correlation is not enough (also see Messing and Mergler, 1995).

Undoubtedly, the epidemiological data are often difficult. Radiation downriver from a nuclear plant (even during decommissioning) might be linked in local people's minds to health hazards. Yet how easy is it to prove a case when there may be, for example, gradual increases in the number of miscarriages, or a small rise in the number of children born with Down's syndrome?[11] The combined government and financial interests involved typically deny any association. Complainants, after all, are often poor and might belong to a disadvantaged group. Nuclear waste in the United States, for example, is usually disposed of on lands belonging to Native Americans—again, with consequences that are denied in terms of miscarriages and congenital ill-health (Parrish, 1994; also see Grossman, 1993). In similar recognition of the racism inherent in the operation of science, several black women pointed out how medical research prioritizes illnesses of the developed world, while ill-health "back home," in the developing countries, is paid little heed.

What has, or might have, happened, to their children was indeed a frequent theme of our conversations with women, for whom "expert opinion" often bears little relationship to human (or animal) lives, or to the real hazards that people face. Many women are (understandably) skeptical about the demand for "scientific proof" when there seems to be plenty of circumstantial evidence of particular risks.[12] They are also well aware of the politics of denial and the inherent racism of science.

Strong ambivalence is also apparent. Despite skepticism, many of us do expect to consult appropriate experts. Some women moved in interviews between skepticism and the need to consult, depending upon the context. Sometimes they comment upon the contradictions that emerge as we speak. This ambivalence toward expert opinion in science is brought out clearly in our interview with Chris (see Box for extract). Chris herself notes the contradiction between her tendency to want to believe factual, scientific reports and her unease at accepting data derived from animal studies, of which she is critical. Despite her recognition of this, however, drawing her attention to it makes her uncertain—"I haven't got my facts right." She shares with many of the women a profound mistrust of experts (in this case, those who use animals in product testing), while simultaneously feeling that they do have some privileged information, to which she does not have access.

BOX ONE

*Extract from interview with Chris:*

*Q: What does the word* experiment *mean to you ?*

*A: Uh, horrific [laughter]. The whole idea of the testing of the animals. I find this unbearable.*

*Q: You put that on your questionnaire, can you explain a bit why?*

*A: Because I think it's unnecessary, a lot of um. . .but you see there again . . . this is going to really . . . my whole belief why I don't like animals being tested is because I don't think they get the correct answers from the animals that they should do, um and then they start and they say we tested five hundred rabbits and this tablet is now safe for consumption . . . [noticing contradiction] I have said all along if someone brought out a scientific research project I would believe it. But I don't think it's necessary, animals are tested far too much on products that have already been tested anyway.*

*Q: You just noticed that there is a bit of contradiction between you saying that scientists produce facts you would believe, and then there's this stuff about animals [affecting what you believe]; could we explore this a bit further, you have just noticed that you have contradicted yourself. So what does that mean to you?*

*A: It means that I haven't got all my facts right doesn't it?*

*Q: No it's what you think about it.*

*A: No well I think maybe I've been brought up to believe a scientific report. Maybe because I have this idea that there's lots of test tubes and bunsen burners and that's what they're doing. . . . I do believe that it's right but when it comes to the side when it's not bunsen burners and test tubes any more, it's animals that's when my whole opinion changes.*

## From experts to the heart

There is more to it, however, than becoming passive in the face of scientific knowledge and expertise. Sometimes, the women actively put something in its place. Thus, in saying "I can know without knowing" (Jenny);

in describing "heart" knowledge as "survival" or "whole" knowledge (Gita); in insisting that "everyone has a natural intuition about what goes on in the world around them" and on the need for science to acknowledge the "nebulous" in things (Selma) and the limitedness of science (Edna)—there is a challenge to the dichotomy of feeling and rationality, as some feminist critics of science have argued (see Sayers, 1983; Rose, 1994). It is, moreover, a challenge in which the prevailing image of science as master narrative is contrasted to an understanding of science as socially and politically constructed (see Barr and Birke, 1994).

In some interviews, metaphors of connection, of listening, of conversation, even, replace those of detachment, observation, control, as more appropriate ways of seeking knowledge of the natural world. Tania claims that feeling at one with nature was one of the most satisfying and important discoveries she had made, yet she makes no connection between this and science as the study of nature. Similarly, Chris, speaking of natural health remedies, says "It's like the earth providing its own answers rather than scientists sitting over their test tubes. I find this comforting." Even students who are themselves in the feminist-inspired Women and Science course, produce such metaphors. Thus, Jenny believes one should "listen to your body, let it tell you what it needs."

Perhaps what is being invoked here is that the modern scientific way of finding out about the natural world is, after all, a recent cultural phenomenon (see Keller, 1992): Carlo Ginzburg's work on historical epistemology documents an alternative scientific paradigm with ancient roots (1980). This "conjectural" tradition is, he argues, rooted in the everyday and the sensual, but not irrational, and is peculiarly the perspective of those who are not in a position of power in a given society.

Often such knowledge is dismissed as trivial or unscientific. "Womanly intuition" is an obvious example—a tricky word which has been denigrated for its association with "mere" feeling and irrationalism but which is more accurately seen as "another way of describing the instantaneous running through of the thought process. . .[and as] neither more nor less than the organ of conjectural knowledge" (Ginzburg, p. 28–29). Some of our interviews show an urgency to apply this central idea to knowledge of the natural world—to acknowledge that natural reality is opaque, that any project to "know" it completely is fantasy, to control it, self-defeating (as feminist critics have often argued). Such discourses fly in the face of the hegemonic discourse of western culture, of scientific certainty that can yield control over nature.

## Knowledge and/in Communities

A central challenge of an education for all women is to confront these divisions and to acknowledge their cultural power. If women are to claim rather than simply receive an education—an act that "can literally mean the difference between life and death" (Rich, 1979, p. 232)—they have to "make visible what has been rendered invisible." This includes, especially, women's work, knowledge, and power (a point argued specifically in relation to science by Hilary Rose, 1994). We need to envisage rationality in less exclusive ways that do not construe emotions and intellect as distinct and separate faculties.

Susanne Langer's strikingly titled book, *Mind: An Essay on Human Feeling* (1988) is helpful here. Her opening sentence is: "Feeling, in the broad sense of whatever is felt in any way, as sensory stimulus or inward tension, pain, emotion or interest, is the mark of mentality." This suggests that the anger with which some defend their "objective knowledge" reveals something about objective knowledge: "it is fraught with emotions and entangled with values that are themselves subjectively felt" (Minnich, 1990, p. 173).

Minnich remains agnostic as to whether or not women come to know differently from men; nonetheless, she thinks it likely that many people think well and effectively in ways that "do not fit" the narrow confines of the dominant culture's notions of rationality and intelligence. Many outsiders want public recognition of other ways of thinking and knowing than those valued and given currency in the mainstream. A narrow view of what is rational has, Minnich believes, created educational systems that can make many of us feel inadequate because "our ways of thinking, of making sense, are not met, recognised, given external form, clarified and then returned to us refined and strengthened" (p. 111). "Knowledge requires many of us," says Minnich. To use a central metaphor of much recent feminist writing, we need to find the "suppressed voices" to change and enrich education.

If science itself is to reflect more fully the experiences of all of us, then those suppressed voices have to become part of the process of creating scientific knowledge. Feminist critics of science have stressed how science needs to include the standpoints of excluded "others" (e.g., Rose, 1994; Harding, 1992).[13] Indeed, if it is to meet its own requirements of objectivity, if it is to represent a "strong" objectivity rather than a weakened form, then it *should* be sensitive to the differing perspectives of all of us.

"Knowledge will never be complete," says Lynn Nelson,

*but the experiences and stories that have in their claim to universality excluded and mystified other experiences and knowledges. . .in reflecting the experiences of privileged men. . .have been partial in terms of what it was or is possible to know in given historical, social and cultural contexts."* (Nelson, 1993, p. 151)

Most of our interviews revealed a consciousness of the exclusion of most people from the central processes and purposes of science. Most also revealed a belief that a more inclusive science would be *better*, not just fairer. This view is expressed most strongly by Gloria, a young black woman suffering from a rare hereditary disease that caused her to be born with no limbs. She believes fervently in science because it has enabled her to live independently. She believes, too, that there is an urgent need for science to become more democratic, in the sense that

*It needs different groups of people—men, women, disabled, elderly, black. . . . You need a good representation of every group in science. Otherwise, how can you help everybody? They'd collect different data because of their different priorities and interests and make-up.*

We cannot, of course, know what difference it would have made to our intellectual heritage if those other knowledges and experiences had been included. Would knowledge that arises out of work within tightly defined abstract systems be highly valued as "objective," as the pinnacle of "real" knowledge? And would what is disparagingly called "women's intuition" be relegated to the bottom of the knowledge hierarchy, as natural and nonintellectual?

Feminists ask these questions not merely to reverse the traditional valuations; rather, they seek to unsettle and destabilize such dualistic ways of thinking so as to make them less regulative and normative. This may help create new conditions for the understanding of difference and hence for the production of new possibilities of "situated knowledge."

Sandra Harding has argued (1992) that learning how to see "from below" (that is, from the standpoint of the lives and experiences of the powerless) offers us the best bet for seeing in a less distorted, less false way. This will enable understanding of the natural world as well as the social world. Harding's version of standpoint epistemology maintains not only that knowledge "from below" will be better or "less false" than knowledge created "from above" but that, "in its capacity to transform ways of looking at and understanding the world, [it] may be instrumental in changing the world we see and live in" (Kuhn, 1994, p. 248).

Magda Gere Lewis has stressed that seeing from below requires more than "offering women spaces within which to speak" or simply including women in the curriculum. Such a strategy does not reach deeply enough into the sources and political potential of women's silence, for one. A central feature of women's collective consciousness is, believes Lewis, the experience of silence. Women are all subject to "those social forces and power relations that would keep them from naming the world from their own experience" (Lewis, 1993, p. 75). All women have experienced exclusion from theory formation. For one of the authors, as for Lewis, not to have noticed during her undergraduate years in studying philosophy in Glasgow that all required reading was by men is something she now feels to have been deeply violating. It was bad enough that there was a dearth of references to Scottish philosophers in a Scottish university—with the exception of David Hume.

Such "soft data" of sexism rankle but are sometimes difficult to describe, particularly individually. Yet what is impossible to achieve as individuals can be done collectively, as we know from experience of the women's movement. We need new pedagogical skills, suggests Lewis, to create curricula out of the invisible and silent (p. 194). This includes learning how to see women's silence as a political act with subversive potential, rather than as indicating deficiency, absence, or lack. Even in academic discourses, ignorance, and its construction, is beginning to emerge as a field of inquiry, alongside knowledge (Smithson, 1989; Stocking and Holstein, 1993).

Shoshana Felman emphasizes that ignorance is not a lack of knowledge, but can be resistance *to* knowledge: the desire to ignore. We can learn from it; its investigation creates a new condition for knowledge (see Felman, 1982; Felman and Laub, 1992; Pagano, 1991). Silence and proclaiming ignorance, whether one's own or someone else's, are thus just as political as claiming knowledge.

This is the standpoint from which we have tried to conduct our research. In listening to women talking about their views of science we listened, too, for the gaps and silences. Our own experience as women, as learners, and as teachers tells us that women's silence is not always an indication simply of an absence of words or a lack; viewed in its political,

historical, and more immediate context (like that of being interviewed by University-based women researchers), women's silence can be a powerfully subversive practice.

There are, then, multifaceted ways of expressing separation from science. Speaking silence is one; mistrust of experts another; insisting on "ignorance" is yet another. Nonetheless, as we have tried to show, such "ignorance" is problematic. For it is not simply a lack; in many ways, it is also about resisting, about distancing oneself. This has been found in several other studies of how members of the lay public perceive science. But for women in general, the issues are deeply intertwined with gender—both in relation to the act of distancing and also to the forms of resistance. Thus, what *is* owned and acknowledged as knowing is frequently referred to in terms redolent of gender (such as womanly intuition). The distancing, too, is overlaid with race and class and the long history of imperialism in which black women learn not "their" science, nor even about their own understandings of nature, but those created by the colonizing power.

Lay people rely on particular, culturally salient (and culturally specific) concepts of nature. These diverge from the concepts of nature described by science. In the next chapter, we turn to some of the ways in which the women articulated their perceptions of nature and the natural, and how they used these expicitly to resist certain kinds of scientific accounts.

## The Limits of Science: Metaphors of Nature

This chapter focuses on the limits of science and descriptions of nature. In many ways, it represents several complex and nuanced strands of thought in which the women began to resist or to circumvent the kind of scientific discourse that, in the context of this interview, might have been expected.

If asked what they would choose to research, had they become scientists (and money were no object), the women consistently refer to

two broad themes: medicine and alleviation of disease and suffering, and environmental issues. An undercurrent to this is often the feeling that scientists may be going too far in some lines of research—that they are going "against nature." For the women here, research should be for the benefit of people (and/or nature) and should be done in harmony with "nature." It is in talking about "nature" that at least some of the women speak most clearly about recognizing their *own* knowledge and its importance, contrasting this with the highly limited and contrived knowledge that scientists produce. Exploring how concepts of nature appear in their conversations is a way of tapping into some of their beliefs about science and about knowledge more generally. Articulating the boundaries of what counts to them as science and/or nature represents a second area of discourse reflecting, we believe, discussions within particular communities.

For example, Afro-Caribbean women tended to emphasize the need for respect for nature and the belief that women are better at conversing with and being respectful toward nature. First, Barbara, a younger Afro-Caribbean woman:

*Men destroy things because they like to feel power. Women are more respectful to nature, perhaps because they give birth. They're less hard. . . . Women preserve day to day. This isn't a priority for men.*

An older woman, belonging to another Afro-Caribbean community group, expresses a similar point:

*Women's caring side gets in the way of hard science and the image of having to divorce yourself from all that's kind and gentle and loving. Girls are put off by its harshness, its lack of humanity. It's not that girls are incapable. They make a choice. (Edna)*

Feminist writers have often noted how "all that's kind and gentle and loving" is expunged from science textbooks, in which military metaphors and examples abound. Such examples pervade areas of science as diverse as ecology (Gross and Averill, 1982) and algebra (Campbell and Campbell-Wright, 1995).

Metaphors of nature, moreover, pepper these women's accounts. Some express feelings that scientists might be over controlling and going "against nature"; others use "nature" to challenge the limited view of science. This kind of nature cannot be fully known by the methods of science and is capable of resisting human attempts at control.

As we saw in chapter 2, the stereotypes of the mad scientist and of science out of control are common, deriving from many parts of western culture. Fears about what scientists might be doing typically center on anxieties about interfering with nature, perhaps thereby creating global catastrophe. Many of the women feel that scientists are too often naive, unaware of the political and moral consequences of their work as they "delve deeper and deeper" into problems (a point with which feminists generally agree).

Beliefs about naturalness can also lie behind moral judgements made about others. Speaking about her ambivalence with regard to treatments for infertility, Sally, from Hillcroft College, feels that

*if you can't have a child, then maybe you weren't meant to have a child. Because I know, this is purely personal, one or two women who have had children in the end with hormones but they are not mothers and I feel the child is going to suffer . . . they shouldn't have been mothers.*

There is clearly a notion of "natural" fertility lurking behind Sally's judgement on these women and their hormone treatments. Women, she feels, can only be proper mothers if they have conceived without medical intervention.

Concerns about what is unnatural centered in these conversations on two areas: genetic engineering and food. We will explore these in some detail here, in order to examine how concepts of nature and the natural played out in these women's discourses. Concerns were expressed, for example, when discussing health. In this context, what is "unnatural" includes food processing and particularly the use of food additives.

Advertising and coverage in the media are important sources of images and information. Reading articles and seeing programs on television about (say) food scares and additives bring discourses about risks and technicalities to a wider audience. These are also important sources for knowledge or beliefs about developments in genetics. At the time of writing, developments in genetics and human reproduction have recently been featured widely in the media. Some of these have adopted a gee-whiz approach, extolling the wonders of science.[14] Most, however, have focused on ethical dilemmas, with the result that there can be few of the adult population now who have not heard of cystic fibrosis or of the idea that sheep can be bred with a "human" gene.

Public opinion surveys in Europe and the United States have shown how the idea that scientists might be dabbling with genes and DNA wor-

ries people. This anxiety is expressed in our interviews. Jo, for example, had recently seen a program on genetic engineering, which worries her because

*you don't know much about it . . . using sheep to produce protein in their milk . . . that's very worrying not knowing enough about it . . . in the program it did show people very concerned about what could happen . . . I think they need, errr, everybody needs to know a bit about science.*

Loretta learned a little about genetics in her "Inside Science" course; she is clearly torn between anxiety about where it was leading and knowledge for its own sake (combined with ambivalence about experts):

*there's got to be a limit, otherwise you could go back to Hitler's way of thinking. I suppose genetic engineering, it's all right for finding out, but whether you'd want to use it in everyday life . . . it does cause a moral dilemma . . . I mean knowledge is very good . . . if they're finding out, there'll always be someone somewhere who it'll help. Maybe a little bit of control, I don't think they should be restricted. If it's morally wrong, maybe in years to come it could be used. But surely scientists know what is right and what is wrong, I would have thought.*

Loretta's optimism about scientists is not shared by many of the women, for whom scientists are not readily to be trusted with moral decisions.

Genetic engineering, many feel, is acceptable only if there are clear and immediate benefits in terms of human health. Thus, Edna feels that research involving the isolation of the gene for muscular dystrophy is important and worthwhile: this is, she says, "science applying itself correctly." On the other hand, Monica is dubious about genetic research, even when it might indirectly lead to health benefits. A group of Hillcroft women were discussing the production of "oncomouse," a strain of mouse carrying a particular gene making it prone to developing tumors, for potential use in cancer research. But, asks Monica, "what is that proving" if they produce a mouse prone to developing cancer? She goes on to say that

*It sounds to me as if they've just been playing around with genes to see what would happen. And it's just a play; they're not even realising about the possible ecological consequences later on. They're just having a play, you know, like children will play . . . And that's what's annoyed me.*

A similar concern is expressed by one of the Muslim participants in our group discussion focusing on Islam (see case study 3, chapter 3), who emphasizes that "a lot of science today is science for science's sake," without concern for moral dilemmas.

For Monica, as for other participants in her group discussion, the need to develop "oncomouse" is bewildering.[15] It is far more important, they concur, for scientists to concentrate on epidemiology and prevention: to address the question, why do we get particular kinds of cancer and what can be done to prevent it? Scientists, as we have noted, are concerned about the public's understanding (and acceptance) of science. The women's views illustrate the extent to which people may simply see the research as missing the point—playing for playing's sake, rather than focusing on real human need. If so, then they will not be likely to "accept" the research unquestioningly.

A related source of worry to many people concerns the creation of transgenic organisms. While this issue came up only once or twice in these interviews, it is something which, in our experience, concerns many people. In teaching lay audiences about new developments in genetics, we have found that a frequent anxiety centers on beliefs about genes carrying an essence of the organism from which they came. Thus, the manufacture of tomatoes carrying genes derived from fish might provoke discussion about how "fishy" the tomatoes would taste.

The case study focusing on Muslim women illustrates this point. Here, the problem is that an animal containing a gene from another species partakes of that other species; consequently, a sheep containing pig genes, say, would flout religious proscriptions against eating meat from pigs. This was clearly a matter of concern to this group, provoking lively discussion.

Geneticists would no doubt scoff at such beliefs. But such anxieties seem to be quite common, reminding us that debate about "public understanding" must take into account people's prior perceptions. In particular, people's understanding of genes is likely to be affected by their beliefs about species, which in turn may be influenced by strong religious beliefs. How we perceive food is affected by cultural constructions of the "natural." Most of those who spoke about natural foods had noticed popular accounts in magazines about what might be good for youngsters. Several women note that food containing additives might be bad for children and might make them hyperactive. At the same time, many

women are skeptical of the claims (while also being anxious about their children's diets).

For Kirsty, for example, "natural" food means "stuff you don't have to open a tin for or a packet. . . . ummm I would say like the traditional [British] Sunday roasts, it's proper food, it's not out of the freezer or what." "Proper" food is thus something that someone has to work to prepare. In some ways, this is the antithesis of the kinds of changes in food production and consumption that have characterized industrialized societies in recent decades (including that of meat: see Fiddes, 1991). As Goodman and Redclift (1991) have pointed out, there have been dramatic shifts since the Second World War in processes of food consumption, with increasing reliance on processed food and take-out food from non-European culinary traditions. These changes in turn are linked to a transfer of domestic tasks to outside specialists (the producers of convenience foods).

Other studies, of working-class women, have suggested that women often use the concept of a "proper" meal, to mean one that they have prepared themselves rather than based on "convenience" foods (Charles and Kerr, 1988). The problem with this, Goodman and Redclift note, is that it relies only on what the women *say* they do, rather than on any observation of their behavior. We only have Kirsty's word for it that she thinks "proper" food is important; we have not followed her around in an ethnographic study of what she eats or prepares for others. Nevertheless, it does draw attention to ways in which industrialization of what we eat can be resisted. For many people in Britain and the United States today, "'healthy' eating has become an alternative to the excesses of the 'fordist' diet" (Goodman and Redclift, p. 252). Emphasizing the need to feed their children "proper" food is, for women like Kirsty, partly an inevitable effect of advertising slogans based on concepts of the "natural" (for example, organically produced food as more "natural" [James, 1993]). But it is also a way of resisting the very industries (and their control over food production) that produced the advertising in the first place.

The rhetoric of "healthy eating" is a familiar one to which many women allude. Sally states her belief that bodies are naturally healthy. For her, eating is itself a scientific activity. She describes how she shops, as well as expressing a certain amount of skepticism:

*I think . . . you see a lot of things on a packet and a lot of us don't really know what they mean. If I can avoid buying anything with anything in it I will. . . . My husband would say, you are going slightly over the top, if it's on the shelf it's perfectly safe. I have a brother who is in the food industry and he says, if it's in the food it's ok, but I think that I don't believe that, purely because I think that I realize that something appears to be safe and then of course they do some research and they find some people are allergic to it and they dig deeper . . . so I don't really trust anything and I think things that are pure are much better for you.*

Some women had learned through experience about food allergies, which alerted them to reading packets. Cath's daughter became ill as a result of an allergy to dairy foods and was eventually treated successfully by a naturopath. Cath was not aware of food allergies until that experience, but afterwards she noted "all the things in the papers about yellow coloring and E numbers and you actually start looking on the back of food packets." She comments on how much awareness of these issues is growing among people: "so it's an education . . . when I think when I was a kid or when I visit where I was born they will sit in a pub and talk about food colorings."

Yet it is precisely around concerns with health that beliefs about nature's separation from culture have been most disrupted, through recent "food scares." Perhaps the most publicized was the existence of Bovine Spongiform Encephalopathy (BSE), or "mad cow disease," found in British cattle herds during the 1980s and causing a kind of dementia in the affected animals.[16] As Allison James (1993) has pointed out, BSE "represented to humankind their own animality as part of the food chain and their vulnerability to the ravages of nature. . . . Something had to be done, and quickly; culture had temporarily lost control over nature. . . . It starkly repositioned humans as part of the natural world in a chain of consumption—cows eat the brains of scrapie-infected dead sheep and humans eat the cows. Nature was threatening to pollute culture, as culture had earlier polluted nature by rupturing the boundaries between herbivore and carnivore, between life and death itself" (p. 214).

Newspapers carried stories of the way British cattle were sickening and of the boycott of British beef by other countries. The food industry responded: the Minister for Agriculture appeared on television extolling the virtues—and safety to the consumer—of British meat, while advertising and supermarkets emphasized organically produced foods as something "you can trust" (James, 1993). Even when scientists did claim to find evidence linking BSE to the development of certain brain disorders in

humans, and several countries had responded by banning the import of British beef, the British Government still insisted that homegrown beef could be trusted.

Advertising for "natural remedies" also suggests that these are things a person can trust not to disrupt her or his body. Although at times, "natural" seems in interviews to convey something that is less manufactured, or that is counterposed to scientific, at other times, "natural remedies" are discussed in terms of their *being* scientific because they are tried and tested. Thus, Barbara, from the black community group Osaba, feels that natural remedies must be scientific

> *because they have had to go through a scientific process to find that they work on illnesses or ailments, so to a degree it is scientific even though it's not scientific . . . because it's natural resources whereas probably the other type of medicine is more man-made.*

Some of the black women refer to a growth of interest in "alternative" healing in their own communities, partly because these systems are "more natural," based on medicines derived "from the earth." For older women, this is based on experience in their countries of origin. As Martha observes of her childhood in Ghana, people felt that indigenous systems of medicine gave them more control than the scientific medicine that, anyway, they could not afford.[17] For her their own system is not "alternative"—"we just know it as herbs and drugs . . . that you use for healing."

Black women are more likely to point to issues around science and imperialism, noting that women in developing countries rarely benefit from scientific knowledge (see also Third World Network, 1993). While wanting to emphasize the importance of their own cultures and traditions, black women are nonetheless scornful of what they see as a western fashion for alternative health. This expresses a specific type of resistance linked to an awareness of the part western science has played in imperialism. Thus, one woman who came to Britain from Zimbabwe, feels that,

> *That's one thing I think you western people are crazy about. OK it's important to be careful about what you eat but you can go too far. And if you say to yourself 'I'm thinking positive' you may be dissociating yourself from reality because life isn't always positive. (Brenda)*

Later, she expresses her awareness and disgust at the expropriation of "traditional" kinds of knowledge:

*Science has improved women's health here but not back home. Cancer of the uterus is neglected and the West is doing nothing about AIDS. And experiments done on herbs here don't get fed back to the Third World, another form of exploitation. (Brenda)*

The appeal to "natural" products is double-edged. They know how the knowledges of their own cultures about (say) the medicinal properties of particular plants have been ignored, yet at the same time they acknowledge that western science, too, must sometimes rely on those very same properties and plants repackaged as scientific medicines. They are also more likely to be aware of the cultural relativism of "natural"; what is construed as natural within their cultures of origin may not be "natural" within a culture that is predominantly white and European.

## Nature Resisting

For these women of non-European origin, "nature" serves as a contrast to Eurocentric science. "Nature" can thus be a metaphor of resistance, sometimes explicitly contrasted to the scientific accounts of textbooks. In her study of the metaphors and language with which women describe physiological processes, Emily Martin (in *The Woman in the Body*) notes the power of the "production" metaphor found commonly in textbooks. This metaphor pictures the body and its physiology as factory production, organized hierarchically so that "master" organs can be identified (such as the brain or the pituitary gland). Women's physiology, in these texts, is described primarily in terms of reproduction. Here, Martin notes, the rhetoric emphasizes failure: menstruation and menopause, for example, represent failed reproduction. She notes how often the language describing women's bodily functions conveys a negative image—tales of breakdown of the uterine lining, failure to implant, menopause as loss or lack.

When the women Martin interviewed were asked to describe how they would describe the changes of menarche to a young girl approaching puberty, many of them produced scientific accounts, replete with the metaphors of failure and breakdown. But many did not. Those who did not were predominantly working-class women, whose accounts emphasized instead the experience of menarche as a transition to womanhood—something to be celebrated for that reason. These women, Martin stresses, were unlikely not to know the scientific accounts; indeed, most had encountered them at school and could call on them if pressed. Their choice

of language and metaphor were, she suggests, more to do with a clear *rejection* of the (middle-class) language of textbooks, which portrayed their bodies as failures.

In our interviews, we asked a similar type of question relating to childbirth rather than menarche. This was the first of three questions in which we sought to probe how, or whether, women might use "scientific" explanations. All the women knew that the focus of the interview conversation was science, so we expected that this would predispose them to producing a "scientific" account. But this was not the case, and what was interesting was the contexts in which alternative accounts appeared. Here, too, there seemed to be refusal of the scientific model if it did not fit with other values and ways of seeing the world.

After asking about childbirth, we asked the women to explain how a microwave works, in order to seek responses to a relatively common (but still relatively new) piece of kitchen technology. We followed that with a question asking them to explain why we have seasons. We asked for explanations of three quite different phenomena, so as to probe a range of understandings. As we were soon to discover, these questions revealed an interesting array of very different kinds of responses.

The question about childbirth produced some "scientific" accounts, in which women explained, for example, how the baby would move downward before the birth. But this was often mixed with a more phenomenological account, stressing the experience of giving birth for a woman. Listening to these mixed accounts, it was as though these women knew that a scientific account might be appropriate, given the nature of the interview, but wanted to refuse the starkness of such language. Several, indeed, mentioned that giving birth was *the* central learning experience of their lives.

While childbirth was associated with awe and wonder for some, for a few, it was difficult to talk about. For these women, the question provoked embarrassment and unfinished sentences. What was noticeable however was that those who found it difficult were older, black women.

An obvious explanation for that lies in the power structure of the interview and raises inevitable questions for us as feminists doing research. We, after all, are both white, and representatives of an elite university (our own respective class positions are more ambiguous). No doubt those are important considerations. But finding it difficult to talk also reflects personal histories in which open discussion about women and reproduction was rare, in which reproduction was not on the school curriculum. The women who did find it hard were either products of education in British ex-colonies (based on the highly academic and elite curriculum of the British educational system) or of "special" education for children with disabilities in Britain. In each case, sex education had been omitted from the curriculum. As Gloria, a disabled black woman noted, she had been told nothing about sex and didn't learn about periods until she was ten years old.

Apart from these differences, we could find no relationship between type of explanation and class, age, or the numbers of children a woman had borne, nor even an association with whether a woman had studied science. Thus, two of the women enrolled on the introductory science course for women, gave an experiential account. Sandy began, for instance, by saying that

*there's a life inside you and it's ready to come out now, it's fully developed and when it's right, umm, your body naturally sort of makes a channel for it to come out.*

But the mixed accounts were more common; thus, Veronica, studying literature, begins by saying:

*It is the most wonderful thing. It's very painful but at the end of it you have got something wonderful.*

She then switches to invoking more "scientific" language, noting that the baby

*starts everything in motion, in order for the body [to have] contractions to force its way out, which are painful, which can be eased these days with drugs.*

Selma, also studying literature (and deeply hostile to science, which she saw as interfering with nature), begins with a scientific account, but soon switches to the wonders of birth:

*Just tell them what happens, how the baby forms inside the mother . . . what part forms first . . . how it revolves around, what happens when it emerges through the canal and it pops out through the uterus, umm I could be very explicit about that because I think it's one of the wonders of the world. It's one of the wonders and beauty of nature . . . it never fails to fill me with wonder and awe when I think about [it].*

Here, then, is one way in which the women were appealing to what Hilary Rose has called "heart knowledge,"[7] to an active construction of

other ways of knowing, to set against what they knew of the scientific/medical accounts that say so little about women's real experiences of giving birth.

Thinking about the seasons also invoked "heart knowledge" in a different way. In an interview about science, we had expected that asking about the seasons would elicit from at least some women comment on planetary motion. Yet few produced this kind of answer. There were two, very striking ways in which the conversation turned at this point. First, as we noted above, the black women who had come to Britain from other countries emphasized the colonial heritage of an education in which they learned about the climatic pattern of a temperate country they had never seen. Second, many of the women spoke poetically of the seasons, invoking the myth of Persephone: the earth must rest in order to be reborn.

Alice, for example (see individual case studies, chapter 3), uses the metaphor of dying and rebirth and links it clearly to what she had learned from other women and their own cycles of life and death. In some cases, this imagery is explicitly used in opposition to the scientific explanation (which, presumably, could have been called upon, but was denied). Veronica counterposed her religion to science, saying that: "I am a great Christian, I'm afraid, so although I like science, I am a great believer in God and the seasons are put there for things to grow, things to rest, things to lay dormant and things to start up again." Similarly, Cath likened it to the body's cycles. After querying explicitly whether we wanted a scientific answer; she then proceeded as follows:

*It's part of the earth's cycle, like our bodies have a cycle. The seasons, they are for, she [the earth] has many cycles, the winter is the sleep time, she rests. Spring is like getting up after a fantastic night's sleep, it should be nice and slow, the sun coming up, the end of the spring she feels buoyant and summer she gives birth and the autumn she lets us gather her children up.*

Barbara also invoked similarity between women's cycles and the cycles of nature:

*nature determines the four seasons like the tree for instance it's like you and me, how we change as we go along, and you change to adapt to feelings of the trees and the land . . . I think nature changes because of our surroundings not because of human beings.*

Not only is this metaphor of birth and rebirth very powerful, but we were also struck by the clear sense of this as a resistance to the scientific, planetary, account. As Emily Martin notes in her study, working-class women must have known the scientific accounts of menstruation from school and/or from any experiences with the medical profession. They prefer to use an experiential account in interviews, she suggests, because they are actively rejecting the scientific story and the stark academic language (which is, after all, not *their* language) in which it is typically expressed.

Here, too, these women seemed to be rejecting a particular account. Some were quite explicit about this, asking us, "do you want the scientific answer?" before going on to use different language. This is not ignorance of the science; it is about refusing a kind of language that is sterile, unable to capture the beauty of nature and seasonal change. It is about refusing a form of knowledge that denies the heart. There is surely a lesson here.

Given that so many of the women spoke of the coldness and lack of caring of science, we were not surprised when several specifically mentioned use of animals in science. Chris, from whom we quoted in the box above, was particularly concerned and raised the issue several times. Ethical concern about animal use came up when we asked the women what the word *experiment* meant to them. We intended to probe what they thought "doing experiments" meant, as a practice—but usually this led to discussion about animals. Brenda, for instance, responded at this point by saying,

*Like animals and things—we should respect animal life more than we do, we find that to be a good way of discovering things but at the same time it's harming that life—I don't believe animals are put on this earth to be used as test objects, there are other ways, but perhaps it's just cheaper to use animals and more convenient.*

For Jenny, studying Women and Science, the animal issue had been pivotal in making subject choices at school:

*Biology is the one subject I came . . . top in. And I really loved it . . . but at the same time there was no way I could dissect an animal and I . . . cannot bear the thought of that side of it. You know, I think we know enough without that . . . if an animal dies naturally, okay, we can investigate it.*

She went on to explain that she was opposed to the use of animals, even for drug testing, on the grounds of species differences:

*even if you did experiments on human beings you would still find that that experiment was applicable to that human but not to the*

*next one, so how can we say that that experiment is applicable to a rat. . . . Any experiment on any living thing is just applicable to that . . . and so numerous experiments on animals just collectively say that well, this proportion of animals react in this particular way.*

For her, using animals to test products was not only unethical but senseless and irrelevant.

"Relevance" to women's lives has often been raised in connection with women and science education, alongside the advocacy of "girl friendly" science. No doubt it makes some difference to include more examples drawn from girl's/women's experiences, so that they are not put off (as one of us was put off from studying advanced applied mathematics) doing some subjects because the examples are simply too alien. But is it enough?

There has certainly been more criticism in recent years of approaches to science education that encourage merely the addition of more appropriate examples. This does nothing explicitly to challenge the nature of science itself as abstract, objective, and masculine (Rosser, 1995). If some people refuse scientific explanations because these have no resonance with things that do have meaning in their lives, or refuse to do science because they find the idea of cutting up dead animals disgusting, then pointing to their lack of knowledge or understanding misses the point. Refusing those explanations is not about ignorance. It is about refusing a view of the world that insists we can distance ourselves from nature.

## Negotiated Meanings

Concerns about what is "natural," or otherwise, are part of a growing interest in environmental issues and are often expressed in relation to food. The food industries themselves consistently emphasize naturalness and purity in their advertising, yet they do so even as foodstuffs are becoming increasingly commodified (Goodman and Redclift, 1991, p. 250).

The transformation of concepts of "nature" and the natural is hardly new. The meanings we attach to these terms are culturally contingent and constantly shifting (Thomas, 1983; Merchant, 1980). Here, however, we draw attention to the way in which these women articulated these meanings and the ways in which differences emerged.

In western industrialized societies, women's status is partly that of primary consumer within the home. In that role women can resist the increasing commodification of food. The effectiveness of consumer revolt against increasingly packaged and processed food can be assessed by the scale of the food industry's reaction against it in the subsequent marketing of "pure" food.

Yet that in its turn is helping to shape how we understand the idea of "natural." Butter, for instance, might be advertised as more natural with a photograph of a cow in a field, eating grass (and to what extent is that natural?). As Alice pointed out, some people may feel that "if it's food, it's ok," as though foodstuffs automatically impose less risk just because they are classified as food. "Nature" is undoubtedly a highly contested terrain. For women, the way in which that terrain is contested is likely to be influenced by their perception of food and nutrition and of their role in relation to food preparation.

That is not to say that differences between women are unimportant. As we noted, several of the black women were scornful of what they saw as a western fashion for "natural health," particularly those women who had come to Britain from places such as the West Indies. Social class, too, is a factor, in that several women pointed out how eating more "natural" foods—ironically—often costs more if these are organically grown.

These conversations demonstrate to us, as feminist teachers, the need to understand more about how women come to understand the nature that science purports to study. It is not a nature unmediated by culture (as scientists might have us believe), but one constructed by that culture—and by individuals and their experiences. The meanings of "natural" are themselves a complex product of many cultural influences. Among these influences is advertising, which specifically targets women as primary consumer and purchaser of foodstuffs. One consistent advertising message exhorts women to purchase such and such a product because "it is more natural" and (we must infer) good for "your family's health."

In relation to science, "natural" seemed to stand in these interviews for much that is antagonistic to science. Yet it can also stand for something that has been tested. As Jo stressed, "people tend to think that the natural way has been tried and tested and perhaps they think that the natural way is more reliable than the medical . . . I can understand why people are frightened about drugs these days." This seems to say that science/medicine is progressing too fast for new products to be adequately tested. For several of the women, "natural" products have stood the test of time.

These conversations illustrate the nuances of meanings that the women gave to science or to nature. They emphasize how facile it is to portray women as simply lacking in scientific knowledge. Each of the women, in

different ways, was adopting a complex set of stances toward science as "expert knowledge"; most countered the way such expert knowledges create outsiders by claiming the importance of other knowledges.

In some ways, these interviews emphasize for us the extent to which women do stand outside science and its expertise. The gap seems even wider than we had anticipated, given the women's consistent feeling that "science is all around yet nothing to do with me." Nonetheless, we are also aware of the extent to which the women were negotiating meanings about science/nature—as, indeed, were we in the course of having these conversations. Partly, they were doing so every day: even the act of shopping entails some degree of negotiation of meanings for many women (thinking about what is natural; how that is represented; how to define what is acceptable risk in food ingredients, and so on).

They were also clearly negotiating meanings with their colleagues and friends in the creation of epistemological communities. Discussion about what we might ask them usually preceded our visits. Or, in the case of black women working in particular community groups, it might represent previous conversations about racism in Britain. Yet again, in the case of the group of Muslim women, it springs from their experiences of being both Muslim and women—and, in most cases, Asian in a white non-Muslim culture.

Women were negotiating meanings in the interview itself. Science teaching needs to be able to take such negotiation of meaning into account, to allow for multiple understandings and perspectives. That, as feminist critics have often pointed out, would be a truly radical challenge. Indeed, such openness to different meanings, purposes, and kinds of science would be a step toward creating a democratized science that represents people's interests instead of the interests of corporations and governments. It is in that spirit of radical challenge that we turn, in the next two chapters, to make connections to wider feminist writing on science in particular and on knowledge in general.

I WAS BETTER AT SCIENCE THAN ANYTHING ELSE AT SCHOOL UNTIL THIRD YEAR [EIGHTH GRADE]. BIOLOGY WAS EASIER. BUT I THINK AT 0-LEVEL [AN EXAMINATION TAKEN AT EQUIVALENT OF TENTH GRADE] YOU CAN RELY ON MEMORY . . . AND WITH CHEMISTRY YOU NEED MORE UNDERSTANDING. WITH PHYSICS IT'S PRETTY MUCH ALL PURE UNDERSTANDING. THE MORE DIFFICULT, ABSTRACT, AWE-INSPIRING, THE MORE SCIENTIFIC. ANYONE WHO CAN GET THEIR MIND ROUND THAT IS VERY IMPRESSIVE TO ME. —MONICA

I THOUGHT I WAS REALLY GOOD AT SCIENCE AT SCHOOL AND REALLY LIKED IT. BUT MY TEACHER TOLD ME I WASN'T REALLY GOOD. I REMEMBER BEING VERY DISAPPOINTED AND I JUST GAVE UP. —MOLLY

The conviction of not being capable of studying science may be widespread; it is prevalent in our interviews. The women believed that "real understanding" is the prerogative of special people with special brains. As Edna summed it up: "I just think of scientists being very very bright, very, very—more bright than anybody else." This belief is a product of a science education that secures its continued elite status.

In this chapter, we link the themes raised in (and by) our interviews with these women with wider issues discussed in feminist writing; in particular, we want to connect to some theoretical issues raised by feminist critiques of science. Feminist teaching must begin by acknowledging

women's place as *knowers*. We are concerned about women's exclusion from science and technology; by insisting on the knowledges of those who feel "outside" science, science itself might begin to become more democratized, as Sandra Harding (1992) has forcefully argued.

In this chapter, we consider three themes that connect the women's views with academic feminist ideas. The first is the opposition between owned, "commonsense" knowledge and "other" knowledge. However differently women phrased it, this kind of tension ran through most of the interviews and was frequently linked explicitly to gender. The second theme relates to feminist critiques of science and concepts of "nature." Our third theme builds upon the creation of "epistemological communities"—or, communities in which knowledge is created. We link this to discussions of epistemology by feminist philosophers and ask how these discussions in turn raise questions for feminist pedagogy—especially in women's studies and in science/technology education.

## Having What It Takes, or Owning Knowledge?

One theme was the contrast between "owned" and academic knowledges. Science education and practice now suffer from what Sandra Harding describes as an "ostrich strategy" (1993, p. 6), the persistent refusal to admit other knowledge claims. We believe that this poses particular problems for nonscientifically trained women, partly because it refuses knowledge claims of anyone who is "outside" science and partly because it ignores the ways in which scientific knowledge has historically become cast as "masculine." Rationality and objectivity, likewise, bear that gendered link, as feminist philosophers have noted many times (e.g., Lloyd, 1984; Keller, 1985).

The links between masculinity and science were forged early in the history of modern science. Londa Schiebinger (1989) describes how science and its institutions were, from the early seventeenth century, explicitly linked with gender. The Royal Society of London, for example, excluded women from its membership. By the end of the eighteenth century, a science stripped of metaphysics, poetry, and rhetoric was being championed by philosophers and scientists; literature was banned from science as "feminine," and Goethe's reputation as a poet was said to ruin his reputation as a scientist!

Schiebinger argues that femininity has come to represent a set of values excluded from the practice of science as we now know it. These excluded values may be part of a larger set of values typically attributed to a broader group of "outsiders," for example, black people, who, like women, have been largely barred from the practices of modern science. Simultaneously, these outsiders can themselves become the object of inquiry. Gender, indeed, has informed the practice of science even in the describing and naming of nature, in terms echoing the social beliefs of those who are naming (see Schiebinger, 1993). It thus becomes a "nature" thoroughly inscribed with gender and race.

Recent scholarship on the legacy of imperialism and the workings of "race" in modern science has re-evaluated the causes and conditions of the development of western scientific thought. This work bears out Schiebinger's claim (see, for example, Needham, 1993). While revealing the debt to independent science traditions of other societies like Africa, India, and China, it also suggests that science has been too narrowly defined—precisely to exclude and devalue other forms of scientific thinking because they are not useful to dominant groups in the West (see also Harding, 1991, chap. 7; Harding, 1993, 1994a; Ginzburg, 1980).

In this set of cultural constructions, those considered to have "real intelligence" must inevitably come to reflect those exclusions. One example of these processes at work is illustrated in Valerie Walkerdine's research on mathematics teaching. Since being good at math is generally regarded as a prerequisite for understanding the supposed "queen" of the sciences, physics, this unusual research is relevant to our study. It is unusual because the majority, by far, of studies of math and gender assume from the outset that boys are better than girls and then seek to explain this "fact"—an example of the circular reasoning that pervades much of the psychological literature on sex differences (see Walkerdine, 1989).

Based on close classroom observation of secondary school math teaching, Walkerdine's research recounts a striking tale of how girls who are actually doing well at math are still seen by their teachers as not really having "what it takes." Boys, on the other hand, who are actually doing poorly, are still credited as "having potential, just being lazy." (We wonder if the two women quoted in the epigraphs to this chapter underwent a similar process.)

According to Walkerdine, girls end up in a double bind: no matter what methods they adopt in their pursuit of mathematical knowledge, none appears correct:

*If they are successful, their teachers consider that they produce this success in the wrong way: by being conscientious . . . and hard-working. Successful boys were credited with natural talent and*

*flexibility, the ability to work hard and take risks. . . . Further, teachers tend to think that boys fit the role of "proper learner"—active, challenging, rule-breaking. (Walkerdine, 1989, p. 155)*

It is hardly surprising that when boys at school are asked why they are not doing well, they say it's because they didn't work hard enough. When girls are asked the same question, they say they aren't clever enough (see Grant, 1994). Even when girls do display the attributes of the proper learner, teachers seem to be threatened by such challenges, whereas, suggests Walkerdine, boys' challenges are often elaborated and extended.

Such prohibitions on girls are often emotionally charged, conveyed in a tone of voice as much as in what is explicitly said. As such they act as a force that "unsettles our understanding," suggests the philosopher, Michele le Doeuff, who, as a schoolgirl, came across a work of Kant. "Overwhelmed" by it and wanting more, she asked her teacher where she might get a copy of Kant's *Critique of Pure Reason*: it was, she was told, in word and gesture, "much too hard" for her. She still has not read it (see Le Doeuff, 1991, p. 142–47).

That prohibitions like this continue to unsettle our understanding in adult life, particularly in relation to what are perceived to be the most abstract systems of thought, seems borne out by our opening quotations.

Walkerdine's research fits well with our own, illustrating how women's beliefs about the need to "have what it takes" develop in the classroom. Particular knowledges, and ways of thinking, can thus become gendered: rationality, for example, becomes associated with masculinity, while "irrationality" becomes feminine and is accorded lower status. Feminist philosophers have argued that the limits of reason have indeed been fixed to exclude certain qualities that are then assigned to women (see Gatens, 1991, p. 95). Their claim is that "femininity" is constituted partly by this exclusion:

*Rational knowledge has been construed as a transcending, transformation or control of natural forces and "the feminine" has been associated with what rational knowledge transcends, dominates or leaves behind. (Lloyd, 1984, p. 2)*

This notion of reason as a method of thinking that sheds the nonintellectual and requires rigorous training is culturally specific. Yet it functions in education as a "mystified concept," the entirely obvious and only way of being rational (see Minnich, 1990). Linda Shepherd (1993) suggests that science should benefit by admitting more of these mystified concepts, notions such as intuition and caring. But, she reminds us, these have been explicitly excluded in the ways that scientists have been trained.[18] She cites two articles about science education, written in 1938, that admonished scientists deliberately to denounce emotions and to learn to "think coldly" (Shepherd, p. 51).

In contrast to the prevailing emphasis on cool logic and reason, Carol Gilligan (1982) has argued that women might prefer a "different voice," one that emphasizes warmth, connection, caring, and context. These were undoubtedly themes that were often voiced by interviewees, who almost unanimously believed that women "thought differently" than men.

Any suggestion that women might think differently is contentious (and especially to those women who have entered "masculine domains," such as women scientists), yet Gilligan's work underscores the need to recognize other ways of thinking, ways that diverge from concepts of duty and universal moral principles. This different voice means "positioning and repositioning oneself within a situation until the best course of action comes to suggest itself; but always at points within the situation, for there is no removed, God's eye vantage point" (Code, 1989, p. 165). It is that supposedly objective, God's eye vantage point that is central to scientific rationality.[19]

The value of this kind of feminist "critique of reason" (see also, Belenky et al.'s 1986 study of "women's ways of knowing") lies in its emphasis on how meanings must be negotiated. It draws attention to the fact that "the certainty, clarity and precision claimed for dominant theoretical structures is as illusory as the truth claimed for stereotypes" (see Code, 1989, p. 169). The nature of science like the "nature" of rationality is no more fixed nor uncontestable than the nature of men and women.

Many of the women in our study clearly identified "hard," "rational," scientific logic with masculinity and "softer," "emotional" qualities with femininity. At the same time they acknowledged that actual men and women (including themselves) do not fit neatly into these stereotypes. So the idea that science is masculine—which ran deeply through the interviews—is perceived as "simultaneously self-evident and nonsensical" (Nelson, 1990, p. 179).

Nearly all the women we spoke to emphasized that science would benefit from including more women, because women "think differently." Veronica, a white woman studying literature, suggested that "Men are black and white and women are more shades of grey," while Catherine, in a women and science course, stressed that there is "the male analytical,

logical side of thinking and the female which is intuitive, emotional, feeling, visual, doesn't have a language and obviously to have a whole science you have got to have both sides." She pointed out that in order to do science, a woman has to "be like a man, they've got to think that way to be a scientist . . . you have to be logical but you're still denied being intuitive." Whether or not that is true (and many scientists would argue that intuition is an important part of doing *good* science), what matters here is her perception that "feminine intuition" is likely to be denied if a woman becomes a scientist.

Dominant and gendered notions of "real intelligence" and rationality may also obscure other ways of being rational, if these do not fit dominant ideals. If recognized and given shape in practical social arrangements and pedagogical practices, these other rationalities could challenge dominant systems of thought that exclude so many.

In linking feminist work on (for example) "women's ways of knowing" and science/rationality, we are not saying that one should be more valued than the other. What we want to emphasize is how women outside the academy express these tensions and experience their own knowledge. It is because the dichotomy is so deeply ingrained in western culture that the problems of women's exclusion exist. We need both to ensure that women (including feminists) embrace objectivity and rationality *and* to ensure that scientific knowledge is developed in ways that acknowledge everyone. Those two tasks are intertwined.

## Pure Distillates of Truth? Science and the Feminist Critiques

*I believe that the scientific community exhibits a model or ideal of rational co-operation set within a strict moral order, the whole having no parallel in any other human activity." (Harré, 1986, Introduction)*

*Although the "rationality of science" is supposed to lie in the fact that scientific understanding is the most open to criticism of all understanding, a crucial area for criticism [is] ruled out for consideration: the social arrangements through which scientific understanding is developed. (Addelson, quoted in Nelson, 1990, p. 181)*

Feminist critiques of science have moved from a primary concern with exposing sexist practice and content in specific sciences (typically biology), to including a broader concern with scientific knowledge: how do we come to know what we know? Several commentaries exist that detail the earlier stages of this critique (see, for example, Biology and Gender Group, 1989; Birke, 1986; Hubbard, 1990). We concentrate here on the later, more epistemologically focused phase of the debate because of our own interest in women's relationship to scientific knowledge. We make no pretence to cover the whole of this debate, only those aspects of particular relevance to the project of this book.

The main agenda of the feminist critique of science now includes "a robust attempt to re-vision a defensible feminist concept of objectivity" (Rose, 1994, p. 93). Several strands are involved in this re-visioning. We turn, first, to the notion of scientific objectivity, which feminist critiques seek to displace—the "God's eye" view of science. Second, we look at feminist ideas about constructing better, because more objective, science.

Work in philosophy and social studies of science has focused increasingly on the ways in which knowledge is constructed. There are, it is argued, no theory-free "facts"; knowledge is always created. Science may claim to be the pursuit of truth, merely discovering the laws of nature, yet even scientific knowledge is created by human beings. There is, moreover, always a slack between theories and the evidence supporting them (that is, scientific theories are underdetermined: see Knorr-Cetina and Mulkay, 1983, pp. 4–5)

This being so, we want to examine further the social factors that shape the direction of scientific knowledge (see Addelson, 1983). It is central to recent feminist philosophy of science that gender provides a division in experience deep enough to make a difference in the direction of research and the content of scientific theorizing. It is central, too, to feminist arguments, that science might be developed more democratically, as a mass phenomenon—"not simply as a democratic duty but as a way of making science better" (Nelson, 1990, p. 170).

Yet the belief persists that science is somehow special; that its methods and language, its alleged reliance on the detached, "value-free" pursuit of truth, allow it privileged access to nature. Scientists speaking during British National Science Week seem to believe not only in "truth," but also that only scientists can "name nature." It is not surprising, then, that conventional wisdom has it that the sciences, properly pursued, provide a pure, value-free method of obtaining knowledge about the natural world.

This is objectivity-as-transcendence: the God's eye view of science or "the God trick" (see Haraway, 1991a) which feminist critics believe functions as an ideology. This in turn helps to define membership of the

scientific community and to draw a veil over how social, political, and cultural values and experiences actually shape the development of scientific knowledge.

Writing about how students studying physics learn to accept this view of science, Karen Barad (1995) notes how concepts such as the uncertainty principle in physics, which led historically to much debate and argument, are dealt with. The "one or two lectures" provide none of the history (not to mention the philosophical assumptions) that are critical to ideas of uncertainty in nature. Barad notes that a typical pedagogy in physics entails the teacher instructing students that a particular theory "works"—it must be right. Thus, she points out (using an instructive metaphor),

> *The scientific method is hailed triumphant. It is as if we are to believe that the scientific method serves as a giant distillation column, removing all biases, allowing patient practitioners to collect the pure distillate of truth. There is no agent in this view of the theory construction: Knower and known are distinct—nature has spoken. (Barad, 1995, p. 66)*

No wonder, then, that scientists become so perplexed by those who theorize *about* scientists, including feminists. Barad invokes another colorful metaphor to make this point:

> *Most scientists shake their heads at their diseased nonscientific colleagues stricken with hermeneutic hemorrhages and metastasized multiculturalisms, feeling secure that the inoculation of the scientific method has saved them from such ugly fates . . . [thus] feminist science scholars commit nothing less than blasphemy in insisting that science is not immune from the rational imperative of the incorporation of critical discourse as part of all human endeavors. (p. 70)*

Barad and other feminist authors point out that the "object of knowledge" of science is never pure, unadulterated nature. It is always nature as, for example, a teleological system or nature as a mechanical system or nature as a complex interactive system (see Knorr-Cetina, 1983; Longino, 1990, p. 99; Harding, 1993). Such metaphors unconsciously influence scientific work and are shaped by the values and concerns of the wider cultures within which they are developed. Modern science favors the application of mathematical hypotheses to nature, the use of controlled experiments, and a mechanical model of reality (Needham, 1993, p. 31). While not all scientists adopt this model it is a dominant one lying behind and guiding advancing technology.

Donna Haraway, scientist turned historian of science, has produced sustained critiques of some of the main metaphors that guide scientific inquiry. She documents how gender-influenced metaphors abound, for example, in the extensive use of military metaphors in contemporary immunology (Haraway, 1991b). She and others have pointed out the ways in which sexism and racism intersect in the sciences, particularly primatology, the study of apes and monkeys (see Haraway, 1989; also Stepan and Gilman, 1993).

Haraway does not believe that we can avoid metaphorical thinking in science or elsewhere, only that we should be conscious of how metaphors shape the way we see the world. They can be empowering; they can also be dangerous. According to Haraway, struggles over what will count as rational accounts of the world are struggles over *how to see*. Metaphors can be different; we could, for example, think of the world—the "object of knowledge" of science—as itself an actor or agent. We could begin to see the world not in terms of "discoveries" but as "conversations" with nature and to accept that "we are not in charge of the world" (Haraway, 1991a, pp. 198ff.). Evelyn Fox Keller, too, has commented on how metaphors of engagement and identification can assist our understanding of the world, revealing how notions like "ensouling" and "a feeling for the organism" guided the work of geneticist Barbara McClintock (Keller, 1983).

The contrast between discourses of militarism and control and those of "conversations" or engagements with nature emerged clearly in our interviews. As we saw in the previous chapter, many women see the cycle of the seasons in organic terms, stressing engagement with nature rather than control. This poetic stance is a particular kind of "conversation" that we might have with nature, and it contrasts to the distancing stances of scientific writing (see Latour, 1987). Some women avoided the imagery and language of control: Selma, for instance, was quite explicit about her feeling that science was too controlling, too damaging—it "struck fear into" her. It was precisely that aspect of modern science that helped her to turn toward literature—for her, an entirely different kind of discourse and engagement with the world.

What social studies of science point to is the important truism that nature as an object of scientific knowledge *is* social and cultural. Some feminists suggest that the way nature is "in itself" does not set many constraints on what humans can believe.[20] There are of course empirical constraints on what we can know that are set by our humanness and by the way the world is. Nevertheless, what is developed as knowledge

depends significantly on who defines the questions worth asking in the first place. As several interviewees point out, the "who" generally excludes women.

The rationality of science is supposed to rest on its openness to rational criticism. Yet because scientists are on the whole members of the dominant race and gender, they are unlikely to detect the influence of culturally induced, commonsense racist and sexist assumptions on their work. Indeed, the "God's eye" view of science itself shields them from doing so.

Furthermore, argue feminist critics of science, so long as alternative points of view, values, and social and cultural experiences are not overtly represented within the scientific community, such shared values are very unlikely indeed to be identified as shaping scientific observation or reasoning. Yet to be open to what has been called "genuinely transformative criticism" (Longino, 1990, p. 112) which might enable science to develop new ways of "reading the world," the scientific community cannot remain blind to its own context-bound assumptions and values—including assumptions about what questions are important to ask. Indeed, it is precisely that blindness to context (or what Ruth Hubbard calls "context stripping") that made many of our interviewees so mistrustful of scientific experts.

Sandra Harding (1993) has made a plea to natural scientists and those in science education to open their ears to other ways of knowing in order to achieve better science. She recommends, first, the incorporation of literary criticism and social, historical, and cultural studies, as well as philosophy. These, she suggests, could provide particularly useful modes of thinking and knowing for a critical, self-reflective science.

This requires abandoning narrow notions of "real" science as that which professionally recognized scientists do, while moral, historical, or political questions are relegated to the ethics of science, the history of science and so on—but not to science itself (see also Minnich, 1990). Recent developments in gene technology provide a clear example. The Human Genome Initiative largely funds basic scientific research; a small sum is set aside for ELSI, the Ethical, Legal and Social Implications, for work done by (for example) sociologists and ethicists.

The women we interviewed, like academic feminist critics of science, perceive a need to "open science up," to make it more responsible and responsive, to contextualize it. Only then, they insist, can science and technology begin to represent everyone and to be more accountable. That aim is one with which feminists would agree.

### Responsible Knowing?

Women's experience is vital to the creation of knowledge. This kind of learning was central to the consciousness-raising groups of the Women's Movement. Through sharing, discussing, and analyzing their experiences and daily lives, women developed collective understandings; they (we) produced a view of politics that challenged external definitions of politics and re-envisaged the personal as political. In doing so, many women developed a re-formed view of themselves—as being the sort of people who could think for themselves and create new knowledge; they developed a re-formed view of knowledge, too.

Nelson maintains that to enlarge "what it is possible to know" means recognizing that communities and groups, rather than individuals, are the "primary agents of epistemology, the primary generators and repositories of knowledge." Hers is a philosophical argument for what she terms "epistemological communities" (extending Thomas Kuhn's concept from the philosophy of science) as the primary agents of knowledge. Yet the real agents of this insight were not primarily philosophers and academics. They were the liberation movements (consisting of many groups and communities) of the 1960s and 1970s who staked claims for the legitimacy of "subjugated knowledges." At the same time they exposed the partiality of official knowledge: Whose knowledge? was the historical and social question that dispelled the fantasy of disembodied knowledge and marked a crucial moment in, for example, feminism's history (see Bordo, 1990).

In relation to science, Sandra Harding argues that science, too, must acknowledge and include values from the perspectives of the least advantaged groups in society. Such "outsider" perspectives are required, she believes, if science is to achieve "strong objectivity," not merely "weak objectivity" (Harding, 1986, 1991, 1993). This requirement demands "affirmative action" as a scientific and epistemological goal, not just as a moral and political imperative:

*Social communities, not either individuals or "no-one at all" should be conceptualised as the "knowers" of scientific knowledge claims. (Harding, 1993, p. 18)*

*Science should be seen for what it is: "the name we give to a set of practices and a body of knowledge delineated by a community, not simply defined by the exigencies of logical proof and experimental verification" (Keller, 1985, p. 4).*

There are two important points being made here. The first is that science communities are not the only epistemological communities although they are the certified knowledge-makers, the ones granted "cognitive authority" in our society (Addelson, 1983; Nelson, 1993). This cognitive authority in turn is maintained through community policing of boundaries: Collins (1985) describes, for example, how "core sets" of scientists operate. It is the core group whose opinions come to count in the arbitration of any controversy, while those who are marginalized can have little say.

The second important point is that an individual's knowing depends on some "we"—some group or community—that constructs and shares knowledge and standards of evidence. Paulo Freire (1970) speaks of consciousness as a social activity and insists that to know implies a dialogical situation. I *can* only know what some "we" can know or learn.

Women talking about science expressed commonality through communities of knowers. Yet that simple phrase, while expressing the act of knowing, does not go far enough: Nelson's concept of epistemological communities works better, for it implies an active creation of knowledge. Both the sense of different communities as *having* and *creating* knowledge, we believe, comes out of our conversations with these different women. Both matter.

As several feminist critics of science have stressed, various points of view, values, and cultural and social experiences need to be represented within and outside the scientific community: various subcommunities (including feminist ones) in critical dialogue with one another. Alliances, mergers, and revisions would result from such a changed structure of "cognitive authority" in which no section of the community would be able to claim privileged or special(ist) knowledge (see Longino, 1993). Donna Haraway refers to the feminist model of how to develop scientific knowledge, which must be a "power-sensitive, not pluralist conversation."

That such an approach to scientific knowledge would be fruitful to women (and to science) is, we believe, borne out by our research. Many of the women we interviewed would not describe themselves as feminists, yet echo the feminist literature in regarding struggles over "how to see" as a key aspect of changing science. Many comments in our questionnaire returns mentioned scientists' perceived detachment from the world. One woman, for example, questions the *looking* that is center stage in western science when she depicts a scientist as "a person with the need to look—

either to prove what they believe or to discover the order of things." Another writes: "looks at things, sometimes in such a way that they fail to see anything." Metaphors of control and manipulation abound, almost half of the returns (46 percent) referring explicitly to science's and scientists' isolated, "separate," and abstract(ed) qualities—its/their "dissociation from reality," in the words of one woman. One respondent clearly sees laboratories as *removed* from nature when she writes: "works in a lab far away from nature."

The metaphors of connection that came out in interviews as ways of understanding the world connect to other feminist writings on epistemologies of resistance among women and third world communities (see Martin, 1989; Shiva, 1988). The language itself is a form of opposition, of resistance.

This also appears to endorse the view that the reconstruction of scientific knowledge is inseparable from the reconstruction of ourselves. In educating scientists, the education of the "emotions" may be as important as the education of "mind"—indeed, is inseparable from it (see Jaggar, 1989, p. 148). We have learned from the experience of women's studies/women's education that to change the curriculum is not just to change what we think about. It is to change who we are (see Minnich, 1990).

All knowledge, feminist critics have insisted, is located or "situated"; knowledge is always social, historical, and embodied. For most feminist theorists the only consistent and "responsible" way of making general theoretical points is to be aware that the knower is located somewhere specific. This goes for scientific knowledge as much as any other.

Like the feminist critiques of science, the weight of feminist theorizing in general has shifted from critique of the sexism of intellectual systems (and of the pernicious and pervasive assumption that man is the measure of all things) to a more creative epistemological project. We turn to this now in relation to our own work.

A central question in this project is: what difference would it make if women were central to the creation of knowledge, instead of marginal? This is a difficult question to pose without seeming to suppose that women constitute a unitary category. Adrienne Rich, for example, has had the charge of biological essentialism levelled against her for speaking of women's studies as a laboratory of ideas for "empowering the difference women make," that is, for developing ways of thinking and structures of thought we cannot as yet even imagine.

In the "epistemology debate" within feminism (see Rose, 1994, p. 21) about what constitutes feminist knowledge (which has been underway within academic feminism since the 1980s) some feminist theoreticians are now asserting that we cannot afford *not* to be essentialist if, that is, we are to establish foundations for a new historical project. We must affirm the positive difference women can make to the development of knowledge and life. Women have to be seen not as different from men, but as bringing about different values, priorities, and standards—even if these values are presently constructed in the context of women's experience of subordination and "otherness" (see, for example, Rose 1994; Hart, 1992; Spivak, 1987).

"Difference" is now critical for feminist theorizing. Recognition of common bonds among women, as the "second sex," is the foundation stone for articulating a feminist standpoint. But a crucial condition for the development of feminist standpoint or standpoints has been the recognition, too, of the differences among women, that is, the difference between "Woman" and women. We are, clearly, not the same (see Harding, 1991, 1994b; Spelman, 1988; Lugones and Spelman, 1983).

As we have noted, it was mainly black women, lesbians, and working-class women who insisted that the struggle for equality had to lie in the assertion of difference within feminism. And not just difference *from* (white, middle-class, heterosexual women) but, rather, difference as marking a condition of possibility or potentiality. This is important for our present discussion of the possibility of creating "better"—and not just because more democratic—science.

Such attention to differences need not encourage an individualized, fragmented view of the world, as critics sometimes fear. We are not committed by it to countless, shifting standpoints and endless relativism.[21] The important differences, which are epistemologically significant, are social and political, rather than merely individual (see Bordo, 1990; Hart, 1992; Rose, 1994; Spivak, 1987).

Groups and communities construct knowledge, so that social and political identity are epistemological factors. Issues of "race," class, and cultural difference as well as gender are important. Embodiment matters, and who knows matters. And it is in relationship with others that knowledge is developed and standpoints reached. Since "who knows" is not primarily individuals but groups, feminist knowledge is different from women's knowledge; and women, like feminists, do not constitute one group but many overlapping or transient communities. For these reasons, we have tried in the research to strike a balance between seeking out similarities *and* differences in the stories women tell us about science—across class, "race," and age, for example, as well as between these, and across temporary adult education groups as well as between them.

An important corollary of all of this is that experience, too, is fundamentally social. It is made possible by our membership in social groups and the concepts, standards, and projects that are shared by the communities of which we are members. Our experience of protons, rabbits, sexual harassment, mothering, or social movements depends on public theories and practices. The emotions we experience reflect the forms of life in which we participate. We could not, for example, feel betrayed if there were no social norms of fidelity (see Jaggar, 1989). This is of fundamental importance to our present discussion because the reasoning of science is informed by "passion" as well as "reason" (Sayers, 1983); and in our divided world this means it is informed by the passions, interests, and values of a minority.[22]

In addition, one achieves a standpoint as a member of a group. A feminist standpoint is an achievement born through political struggle in engagement with others involved in the same community of interest and concern. Many groups are transient—as is the case with the groups of women who form the basis of our study—but their importance in the creation of "situated knowledges" should not be underestimated. Nor should the dynamics of the group discussing the potential interview prior to our arrival; in that sense, we as researchers also form part of an epistemological community centered upon the research interview itself.

There is no litmus test for identifying epistemological communities (Nelson, 1993, p. 149). Indeed, these are "multiple, historically contingent, and dynamic . . . have fuzzy, often overlapping boundaries . . . evolve, dissolve and re-combine . . . and have a variety of 'purposes' which may include . . . but frequently do not include (as a priority) the production of knowledge" (Nelson, 1993, p. 125). We think that the identification and examination of epistemological communities provides a dynamic to standpoint epistemology and that fostering a concern with science among a multiplicity of such communities could contribute to a more democratic science.

According to Nelson, no standpoint should be accorded privileged status by virtue of its subject position or social location. And however great the chasm between feminist and nonfeminist views of science, it is not unbridgeable: there is a basis for communication and argument

between us because we are all members of different communities and subcommunities that overlap. The differences in beliefs and experiences between feminists and nonfeminists, between marginal groups and others, are epistemologically significant but they are not global. At present the authority of science rests on the majority being excluded—or included only as underlaborers.[23] It is this that has to change.

Feminism demands a dialogue and discourse among women worldwide that transcends academic and knowledge boundaries (see Harding, 1992). A basic premise of feminist standpoint epistemology is that knowledge or knowledges should be useful to the producers—it should be "really useful," not merely useful. Yet a striking feature of the feminist epistemological debate surrounding science that has been in progress for over a decade is that it has so far been conducted at a highly abstract level, among a small group of feminist women academics and scholars (who are mainly white and North American) who constitute a tiny epistemological community. Lugones and Spelman have commented:

*Theory cannot be useful to anyone interested in resistance and change unless there is reason to believe that knowing what a theory means and believing it to be true have some connection to resistance and change. (Lugones and Spelman, 1983, p. 579)*

For feminists the purpose of "doing epistemology" cannot be to satisfy curiosity alone. A critical test of the recent debate about knowledge within feminism must therefore be its effects on the struggle for "really useful knowledge" occurring in a wider frame of reference than the academy. There will be little progress toward the goal of "really useful," empowering knowledge until feminist epistemological debates are brought down to earth and practical spaces are opened up for democratic knowledge making:

*If we wish to empower diverse voices, we would do better, I believe, to shift strategy [from methodological debate] . . . to the messier, more slippery, practical struggle to create institutions and communities that will not permit some groups of people to make determinations about reality for all. (Bordo, 1990, p. 142)*

It is this messy, slippery, practical struggle that engages us as feminist teachers and researchers in adult education. The question of knowledge that so concerns feminists has to be taken out of the academy in order to grapple with the thoughts, concerns, and feelings of other "epistemological communities." This struggle is particularly crucial in the case of science because of the authority it wields and because it relies on exclusive, expert knowledge, created by epistemological communities that have historically acted to exclude women (among others). It is also particularly important at this moment in the history of radical adult education. For it is a moment when such practical epistemological spaces as *have* been opened up, through struggle, are in grave danger of being lost—from memory as well as from the landscape of adult education. In identifying and opening up such spaces, women and others who have been historically marginalized by powerful knowledges and institutions would stand to benefit. Their experiences could begin to count in the creation and legitimation of knowledge.

## Negotiated Meanings: Imparting Knowledge

*M: . . . there is no place yet for the black woman [in science]. . .you know, our white counterparts . . . perpetuated this myth about the black woman as matriarch . . . which is a powerful statement from [one] woman to another, you know I am telling you, you are white I am black . . . and when you have got all that going on, when the [black] woman actually decides that she wants to put pen to paper and say, right, this is how I see something will go, she's going to be sent down each time before she can make a scientist, before she can make any strides forward, and it's not going to happen by men specifically it's going to be done, umm, by other women.*

*I: . . . If you were a scientist, what would your priorities be?*

*M: Imparting my knowledge, to at least one other female. (extracts from interview with Monica)*

Negotiating meanings, including different perspectives, and democratizing science—these are radical agendas indeed. As Monica suggests, it is not only the oppression of women with which women have to contend, but also the ways in which racism can act to marginalize nonwhite women. Yet, in her interview, there is an optimistic note, as she recognizes the importance of sharing knowledge. In sharing knowledge, in different communities, we help to create it.

In this final chapter, we want first to draw out some of the implications of our research findings for sharing and creating knowledge, for radical challenges. The first thing to note is the remarkable similarity between the women's voices in our research and feminist critiques of science. The second is the potential for the enrichment of these critiques by the inclusion of many more voices in the conversation. We want to stress here that if feminists are to extend the conversation to include many subjects—the dialogue among women worldwide that feminism demands—academic and knowledge boundaries must be transcended.

## Educating Women: (Re)creating Knowledge

The recognition that mainstream education offers a partial view stimulated the growth of women's studies/women's education. In the United States the strongest women's studies networks were developed in higher education where women's studies has been carried openly into a challenge to all received bodies of knowledge (Boxer, 1988). Important though this development is, its existence in universities is ambiguous; it cannot radically challenge established patterns of education, nor can it meet the needs of the majority of women. In Britain, by contrast, women's studies initially developed largely outside universities.

Adult education in the United States is altogether differently located—historically, conceptually, and institutionally—from adult education in the United Kingdom. In the United States, adult education is conceived as equivalent to the "education of adults" and the literature on adult education is dominated by notions of "self-directed learning" (see Knowles, 1978). The notions of an adult education *movement* and of collective learning are largely absent from the documented history of adult educa-

tion in the United States. In the United Kingdom, in contrast, there is greater awareness—among various publics as well as among adult educators themselves—of a historical tradition of adult education bound to the emergence and struggles of working-class (and, more recently, feminist) movements and collective organizations like trade unions.

From its beginnings as a social movement to extend educational opportunities, adult education has had both a liberal and radical wing. Its discourse has come to recognize the essentially contested nature of education, oscillating between representations of education as personal development and those of empowerment. Over the years, it has given birth to myriad epistemological communities acquiring and developing new knowledge and carving out new areas of study—including women's studies.

Learning through active engagement with groups has been central to feminist pedagogic practice. Most groups were general in political aims; others focused on specialist areas such as health—or science (see Brighton Women and Science Group, 1980). Self-help health groups, in particular, produced knowledge about medical phenomena different from that provided by medical science and "experts"—knowledge and skills which drew on women's own experiences and needs (see McNeil, 1987; Bell, 1994). They, alongside environmental groups and third world science movements, illustrate the active participation of lay people in constructing really useful knowledge around science (see Shiva, 1988; Braidotti et al., 1994).

However, some of these campaigning groups have now become professionalized, using "expert" scientific knowledge. In the case of women's health groups, for example, Wellwomen clinics have developed, replacing self-help groups. This change is double-edged. More women now have access to better health care but the women involved have also become "patients." In this way politicized understandings have become translated into objects of state intervention and spaces for challenging, oppositional knowledge of the world have been reduced.

Feminist *political* discussion is increasingly being replaced by psychological diagnosis as a way of understanding the world; this is an aspect of a more general cultural phenomenon in which psychological ways of thinking are replacing political discussion: "Therapy is simply a form of translation from one language (the language of politics) to another (the language of psychological health and sickness)" (Kitzinger and Perkins, 1993, p. 95). The price could be high, as "the feminist goal of changing the world is displaced by the therapeutic goal of changing ourselves." In place of "the personal is political" the "personalising of the political is rife" (p. 189). The rich tapestry of women's education described above is similarly in tatters. It is gradually being replaced by—on the one hand, assertiveness training, aromatherapy, and self-development courses—and on the other, by narrowly based vocational training.

Yet current social and cultural realities offer new opportunities for a more radical agenda for adult education. Many of the dramatic changes underway in civil society give rise to many new educational and cultural demands; many of these are scarcely heard and barely articulated. The radical restructuring of work worldwide; the "housewifization" of labor (see Hart, 1992); the increase in global nonpaid working time; the development of a society from which more and more social groups feel excluded; problems of the environment; the growing health gap as well as scientific agendas from which the majority of the population are excluded. All of these demand educational strategies based on social participation and the empowerment of citizens, strategies, that is, which are geared to increasing social and cultural engagement and creativity (see Belanger, 1994).

Such visions fit well with calls to democratize science and science education. Much of the feminist critique of science and knowledge assumes such democratization. If we make these changes, science would be better because it would then represent the voices and knowledges of a wider range of people (Nelson, 1990, p. 170; see also Harding, 1991). Yet, apart from groups focusing on health, women's education groups in the community and women's studies in both the academy and adult education courses have largely ignored science. Moreover, there remains a strong tendency in the epistemological community of academic women's studies simply to see science as heavily patriarchal, rather than to analyze it in detail and deconstruct its conceptual frameworks. As a result, science and technology are largely absent from much women's studies teaching and research. Even within the elite communities of the academy, this absence serves only to reinforce women's exclusion from science and its knowledges.

### Sea Changes: Making Women's Knowledge Count

If feminism *is* to achieve a worldwide dialogue about science, the barriers preventing some women from speaking for themselves need to shift. From this point of view it matters a great deal who has set the terms of the conversation in the first place.

In our own research, the kind of conversations we had with the women

was conditioned in many subtle ways by our own social position as white women academics. Women sometimes made explicit reference to that difference: Monica, in the quotation above, for example, felt able to speak about racism and the fact of the interviewer's whiteness. We acknowledge that our whiteness is a significant factor in setting the terms of the discussion; whiteness, as Toni Morrison has stressed, is a concept assumed and deeply entrenched in western literature and culture (Morrison, 1994). We would want to emphasize that not only should a more radical science education pay more heed to other ways of making science, and to its own racism, but it should also attend to the ways whiteness itself remains unproblematized in scientific discourses.

If women worldwide are to engage in and with feminist thinking about science and to develop theory (knowledge in and about science) jointly, how should this be arranged? And how can feminist scholarship inform and assist practical struggles around science, gender, and race—for example, in struggles to do with environmental damage in developing countries? Feminist theorists in the academy have produced very radical ideas about science, but they have often done so within traditional modes of scholarly discourse.

In chapter 3 we illustrated how shared frameworks may develop even in transient groups like adult education communities. We suggested that such differences in adult learning experiences may contribute to difference in perceptions of science. Gender, race, class, and age are important epistemological factors; they shape our knowing (as Sandra Harding [1991] has argued in her feminist analyses of philosophy of science). And it is because social communities are logically prior to individuals as agents of knowledge that class, race, and gender *are* epistemological variables.

A critical science education for women—for anyone—would accord these factors a more active epistemological status than is common in our education system. A critical science education that took them seriously would not only make women familiar with "the very serious game of the production of scientific knowledge" (Larochelles and Desantels, 1991, p. 387), but it would also encourage them to see themselves as implicated in this, as "responsible knowers" (Haraway, 1991b, p. 107; and Code, 1989) and creators of knowledge. For, as Harding (1992) would have it, we are all inside science.

We have stressed that feminist critics of science uphold the possibility of developing science more democratically, as a mass phenomenon, with the clear implication that there must be radical changes in education. It seems clear to us that an approach to science education that is based on the idea of "responsible knowing" would be highly appropriate to a number of the adult women involved in our research, for whom contingency and uncertainty are not difficult notions to grasp, and yet whose experiences and knowledge have been so often disregarded.

We referred earlier to a number of tensions and contradictions that appear to permeate the research. We believe these say something about women's shared social position in relation to scientific knowledge. The first clear pedagogical implication of this is that if the "science curriculum" is taught to women as it is this will perpetuate whatever created the tensions in the first place. The contradictions emerge out of women's marginalization from the authority of scientific knowledge while they simultaneously recognize that "science is everywhere." A science curriculum that ignores these tensions plays into the gendered and racist structures that created the marginalization in the first place. A greater openness is required for different "ways of knowing," which can be potential resources in the development of new knowledge of the natural world. Knowing the facts is not enough in this radically different approach; knowing how those "facts" were developed, by whom, and in what context is equally important.

The second implication is that if women are to claim and create knowledge rather than simply receive it, "starting from where women are" must be the most useful standpoint. This might mean, for example, encouraging them to label knowledge they already have as scientific and, as such, subject to empirical check like any other scientific knowledge. It must also mean valuing and validating the knowledge that they bring from their membership in their different communities. The women who had come to live in Britain from Africa, for instance, brought experience and knowledge of healing and of the natural world that must be valued, not seen as "other" (or worse, as pseudoscience or quackery).

## Building on experience

Broadening the meaning(s) of science and science education to encompass a wide range of people's experiences, as feminist critics have urged, is necessary if the perception that science is done by experts and "has nothing to do with me" is not to be reinforced. Although we should have reservations about basing an education on women's "different ways of knowing," an equal education for women of all social groups (as for the

men of unprivileged groups) cannot be the same as the education that has been developed in a culture based on the exclusion of some of these groups (Minnich, 1990, p. 109).

Furthermore, to be critical and sensitive to lived experience it is not enough to tack on a bit of history (the history of the great women of science, for example) to existing curricula. For adult women—who have already become outsiders to science—that approach simply reinforces the perception that science is done by experts and has "nothing to do with me." "Girl friendly" science, as this has been encouraged in the school curriculum, is often similarly inappropriate and patronizing. If we want to integrate into science education conscious reflection on what is involved in the production of scientific knowledge—which is a central motif of this book—then a more discursive, consciousness-raising approach is needed. Such an approach might create resistance in some circles; we are familiar with the refrain that a curriculum that was built on women's experiences might "not be real science any more." We would suggest that, indeed, it might become better science because it is more conducive to what Sandra Harding refers to as "strong objectivity"—rather than the weaker form in which "objectivity" stands for the position of a select minority (Harding, 1992).

For many of the women in our study, being in women-only space was important. As one of the women attending the Women and Science course pointed out,

> *I decided that when I came on this course, because it was all women, that I felt safe. If I had applied to a university to do a science course, I mean I don't have any qualifications so I don't know whether I would have been considered for a start, but ummm if I had, if I could have got in there I have decided now that I wouldn't want to be there because I feel that the minute I go in I am going to be restricted because I can't do this or that . . . you know, you have got to follow this path that they've laid down for me. (Jenny)*

Harding also suggests various ways in which a multicultural science might challenge the hegemonic belief that science is intrinsically western. Part of this entails recognizing and validating the importance of science from other cultures. Part, too, entails relocating science projects "on the more accurate historical map created by the new postcolonial studies, instead of on the familiar one charted by Eurocentric accounts" (1994a, p. 327). She asks us to imagine a science department educating its students in, for example, "the role of Biology, Chemistry and Physics in the modern European Empire—and vice versa." Challenging and changing what is taught as "science" in schools and university departments is indeed part of the struggle facing those who seek to democratize science.

Moves such as these would broaden the meaning(s) of science and science education to encompass a wider range of people's experiences, as feminist critics have urged. But they might also "re-enchant" science, make it more fun, because more accessible and inclusive. That science might be fun is a theme that has largely been omitted from feminist accounts (not surprisingly, as we have concentrated effort on criticizing it). Moheno (1993) has urged the need to develop forms of science education that counter the modern "disenchantment of the world" (and science) by contextualizing science. Such education, he points out, would not only be fun, but would help to empower people positively to change the world. Having fun, empowering people, and building on experience are notions that would transform science education. They are not themes recalled by any of the women we spoke to, whatever their age or background.

To build on experience means moving away from the prevailing model of scientific knowledge as facts and certainty. Narrow notions of reason and "science" deny us rich possibilities, as does labelling as irrational anything to do with the emotions, experience, or intuition. Part of the problem posed by science (and the distance of most of us from it) is that separation of feelings and reason. Therefore, one way of developing a more feminist approach to science rooted in women's experiences would be to bring those experiences consciously into teaching and learning. We could share, for example, what it feels like to learn about (say) genetic disease as part of the process of learning/teaching about genetics. We should be prepared to listen if people express (say) fear of electricity, or anxiety that moving genes around threatens humanity. It is all too easy for scientists (or other academics) to disparage such fears: it is much harder to listen and to try to understand why people believe what they do.

This is a far cry from current ways of teaching science, from which feelings and uncertainty are expunged. French (1989) notes how, in science classrooms, teachers' speech itself tends to suggest invariance and certainty; the sought outcome, she points out, is the "normal or unmarked case" (p. 19)—no explanation is given as to why, say, heating a compound may lead to production of a gas or how we know that the gas is oxygen. Any other result, of course, is simply "wrong." Not only is the experience of being wrong liable to be offputting, but it can also ignore prior knowledge. One example is that given by Russell and Munby (1989), in which

a science teacher tried to put across the idea of two kingdoms in the living world, plants and animals. The student who resisted, and wanted to classify instead on the basis of people versus animals, was simply ignored by the teacher. No doubt she preferred her own common sense.

Through integrating experience, the notion of science as "boundary-less" (Nelson, 1990, p. 11), and as "all around us" becomes more tangible. That is, it can become inseparable from common sense, politics, philosophy, history, language, and metaphor and so less exclusive, more human, and more "ownable" by women.[24] As we noted previously, quoting Ruth Hubbard (1990), it is as political beings that women will change science—that is, as citizens.

A critical science education would, in consequence, involve working with women's groups in the community, drawing on their own agendas, whether to do with housing, health, roads or the environment, in an effort to develop more broadly based "scientific communities." The kind of science education we envisage here is an aspect of citizenship education, of "pedagogy through politics" rather than a pedagogy centered solely on the classroom (see Le Doeuff, 1991).

## Articulating Science

An important perspective on adults and science comes from a study by David Layton and colleagues (1993), tellingly entitled *Inarticulate Science?*. After discussing several case studies of adults' interaction with science, they stress how adults *use* science in complex ways; for example,

> *importance was given to the source of the science, and particularly to the extent to which it could be judged trustworthy and reflective of understanding of their situation. Emotion, social relationships and social structures all played a significant part in determining the course of practical action . . . the process of "coming to know science" was inescapably a social one . . . scientific knowledge becomes as much a resource for the construction and maintenance of personal identity, a sense of "who they are," as an external instrumentality for understanding and manipulation of the material world.*

Among other things, these researchers urge, developing science for all must involve consideration of both universal *and* local knowledge.

Our research further underlines the central importance of the group and its dynamics in learning; this may be groups based on the locality; it may be groups brought together because of some shared adversity (environmental or health-related, for example); it may be groups sharing particular perspectives because of religious or other worldviews (such as the Muslim women); or it may be more transient groups such as people in a class. It is surely a challenge for those involved in teaching and working with community groups to be able to recognize and build upon these multiple situations and use them to enrich the experience of working with science.

We also wish to stress the need for a critical science education based on recent curricular developments in women's adult education. From this work we know some of the processes that encourage collective knowledge making. We know from it, too, that a purely intellectual approach to science is not enough, that what is needed is a large number of people in movement seeking change—a collective effort to develop really useful knowledge of the world in view of a future society.

Mechtild Hart's questions are important here. What *would* a population that is concerned about preserving the natural conditions of life have to know? What would it have to learn? And what skills, competencies, and attitudes would allow for understanding and knowing nature in a noncontrolling way? (Hart, 1992, p. 203). Questions such as these urge the development of a kind of knowledge (or knowledges) which, in referring to the future rather than the past, would not reproduce the "science curriculum" but would, on the contrary, transform it. All learning, all knowledge is ideological. Sometimes this is explicit, sometimes not. Adult education can play a role in providing a space for such "knowledge from below." The women in our study could make a real contribution to such a project, and the further identification and exploration of epistemological communities that might take part in such a research/education project seem justified. We also need to be open to thinking about less obvious ways of knowing. Women, for example, may learn some science from their children, or through involvement in mother and child groups. Joan Solomon (1992) has noted how Asian mothers, particularly, worked with their children on a home science project, thus learning some science in the process; white women were more likely to shy away from doing science. As she points out, "The [Asian] mothers who understood what was involved were clearly interested and saw no reason to feel that the subject was 'only for men.'" She goes on to quote from an Asian community worker who felt that, "Provided that it is the mother . . . you aim at, there is a chance of making a real difference in their lives" (p. 13).

On a global level, there are some challenges to the status quo. The

Dutch "Science Shops," mentioned in chapter 1, are an example (in Britain, there are now two places offering similar services). These act as brokers between community groups, seeking help to solve a particular problem (how to deal with local pollution, for instance), and university researchers. What is important about these initiatives is that the questions that guide research come from the community, not from researchers. There are, too, "people's science" movements in India and Africa (Layton et al., 1993). One theme of the international meeting of the organization Gender and Science and Technology (GASAT), held in India in 1996, was bringing science to rural women and girls.

Initiatives such as these are urgent. Not only are women (and many others) marginalized by the master narrative of science, we may be moving toward new mechanisms of knowledge production that could further enhance the marginalization of marginalized groups. As we have noted earlier, Ulrich Beck (1992) emphasizes how ordinary people are increasingly beset by risks that they cannot see and over which they have no control. In the "risk society," the knowledge created by "experts" has somber significance; it can save life or kill. Knowledge production, moreover, is becoming institutionalized in new ways. Michael Gibbons and colleagues have argued that the production of knowledge is now characterized by greater transdisciplinarity; it is produced in a wider arena, including government think tanks and commercial organizations; and it is critically dependent on global communications and electronics. It encourages scientists, among others, to become less interested in solving basic problems and more interested in the market (the fascination of many scientists with biotechnology is an example). This kind of knowledge production is, they argue, more distributed and open ended—but is also likely to enhance global inequalities (Gibbons et al., 1994). Working-class women, for instance, are highly unlikely to be participating in this emerging nexus of knowledge creation linked to markets (except as nimble fingers for the manufacture of electronic components in sweatshop factories in Asia).

We urgently need to rethink our practices as adult educators in ways which are in keeping with changed social and cultural realities, both locally and globally. We have stressed the role of metaphor in shaping our perceptions of the world. Metaphors can have empowering potential and can help free us from ways of thinking that limit and restrict. In groping our way forward we might do well to cultivate a kind of thinking that is beautifully conveyed in Hannah Arendt's metaphor of the sea diver. This form of thinking, though located in the past, is directed at the future:

*This thinking, fed by the present [may work] with the "thought fragments" it can wrest from the past and gather about itself. Like a pearl diver who descends to the bottom of the sea . . . to pry loose . . . the pearls and corals in the depths . . . this thinking delves into the depths of the past—but not to resuscitate it the way it was. What guides this thinking is the conviction that . . . in the depth of the sea, into which sinks and dissolves what once was alive, some things "suffer a sea-change" . . . as though they waited only for the pearl-diver. (Arendt, 1968, pp. 50–51)*

There are several pearls to be prized from the recent past of women's education. Our own research provides reasons for optimism. But not only must we hold onto, pry loose, and develop those radical traditions in adult education but we must seek to develop them further to challenge the *most* powerful systems of knowledge invented by "man."

Women and the many excluded others would stand to gain. Their experiences could begin to count in the creation and legitimation of knowledge. And in the process, science education stands to lose its "ostrich strategy," its persistent refusal to admit other knowledge claims (Harding, 1993, p. 6).

Nobody has a monopoly on knowledge. Further, as Said has pointed out,

*The fact is, we are mixed in with one another in ways that most national systems of education have not dreamed of. . . . To match knowledge in the Arts and Sciences with these integrative realities . . . is the intellectual and cultural challenge of the moment. (Said, 1993, p. 401)*

The persistence of the "two cultures" in education is particularly disabling in this respect. How often is science brought into, say, literature courses, or into "return to learn" courses designed for women? And how often are literature or history brought into science? We can make science seem less divorced from everyday life by breaking down these barriers; we could integrate philosophical, historical, and sociocultural approaches as well as attention to literary and metaphorical devices used in the sciences.

One concrete example is our own work with local women artists; the project "Women, Art, and Science" used the arts and sciences to ask questions about the world around us. Art introduced science around themes interesting to women. Breaking down boundaries in such ways would be attractive to many of the women in our study.

What we have argued in this book, however, is that difference needs

space, too, for the development of new possibilities of being and knowing. These need to grow apace with dreams of integration and commonality—the "dream of a common language." We need to explore and develop a richer range of lives and voices for the "human" common education of the sort Said dreams. We *are* mixed in with one another in ways that education systems have scarcely acknowledged, but we are not all the same. Unity calls on us to remember our human connectedness, but it can be dangerous when we misconstrue sameness.

Some of these themes are central to feminist approaches to education. But feminists tend to stand outside science: if we are to become responsible knowers, feminists, and educators, then we cannot afford to do so. Standing outside science, adopting an antiscience stance as some feminist writing does, is simply not an option: we are all "inside" science; it affects us all.

Donna Haraway (1991) argues this point eloquently:

*To ignore, to fail to engage in the process of making science, and to attend only to the use and abuse of the results of scientific work is irresponsible. I believe it is even less responsible in present historical conditions to pursue anti-scientific tales about nature that idealize women, nurturing, or some other entity argued to be free of male war-tained pollution. Scientific stories have too much power as public myth to effect meaning in our lives. (p. 107)*

While we must all take responsibility for creating knowledge, teachers have a particular responsibility. "The academy," notes bell hooks (1994), "is not paradise. But learning is a place where paradise can be created. The classroom, with all its limitations, remains a location of possibility . . . we [can] collectively imagine ways to move beyond boundaries, to transgress. This is education as the practice of freedom" (p. 207).

To move beyond boundaries: that is truly engaging in the process of making science. And it is a pearl beyond price.

# Appendix

## The Interviews

The research began with a pilot study based on interviews with 15 women who had attended a basic adult education course—"Inside Science"—for women (Birke, 1992). These women ranged in age from 22 to 60; all were white (as were all the women who attended the course), and most were working class.

The main part of the research on which this book is based, however, consisted of three phases. We sought to sample from a range of women attending different kinds of courses for adults (not necessarily science-based courses) or participating in community groups. Most of these were in the West Midlands town of Coventry. The economy of this town has depended to a large extent on the automobile industry, combined with light industry. It is fairly diverse ethnically, with substantial groups of Afro-Caribbean people and Asians (both Hindus and Muslims, as well as some people from China and Southeast Asia). We also interviewed some women who were studying at residential colleges for women or with a rural health group in Derbyshire.

In the first phase of this project, we approached these various groups and asked the women to fill out a questionnaire: 110 women did so. The questions were made deliberately open-ended because we wanted to find out what images of science sprang to mind.[1] We asked them, for example, to finish the sentence, "My image of science is. . . ."

In the second phase, we conducted semi-structured in-depth interviews with 40 women drawn from the 110 who had completed questionnaires. For this stage of the research we concentrated on women from a narrower range of courses, institutions, and groups: one science-based women's

course, an institution offering science-related Access courses (e.g. Access to Nursing),[2] a residential women's college, a community-based rural health group, two black women's community groups, and a literary and cultural studies course. Having selected these groups for the kind of spread they gave us, we then sought volunteers from these groups for interview.

The ages of the women ranged from 23 to 66, with a median age of 42. Of the 40 women who volunteered to be interviewed, 8 were black, 2 were Asian, and 30 were white. Social class analysis is notoriously slippery where women are concerned. According to father's occupation (and using the model of British class structure found in Goldthorpe et al. [1987]), just over half of the women were "working class" (semi-skilled and unskilled manual—occupational class vii); the remaining fathers had mainly lower professional and administrative "middle class" occupations (occupational class ii). According to the women's own occupational position (current or last job) the picture changes, with a large majority falling more or less equally into routine non-manual positions, mainly clerical/shop assistants (occupational class iii) or semi-skilled manual ones (occupational class vii). A minority were in lower professional and administrative jobs (occupational class iii)—mainly nursing or school teaching. Part-time and sporadic employment and taking jobs at a lower level on returning to work after having children were features of many of the women's working lives.

Our prior analysis of the questionnaires helped give focus to the terms of the interviews, which consisted in extended conversations about the meaning of science and scientific knowledge to the women involved. At the beginning of each interview we stressed that our interest was in how the women felt about and perceived science, scientists, and scientific knowledge, and we made it clear that there were no right or wrong answers. The interviews were semi-structured—more like conversations with some, but following a question-and-answer sequence with women who were more comfortable with this format. The same ground was covered as in the questionnaires, but now in greater depth and with additional questions and discussion topics. Additional topics included important learning experiences, preferred ways of understanding various phenomena (such as childbirth), and perceptions of alternative health care. Interviews lasted from one hour to two hours. They were tape-recorded and, immediately afterwards, roughly transcribed by the interviewer (either JB or LB). Notes were taken on the emotional atmosphere and other non-discursive aspects of the interviews, and interviews were later transcribed in full by a professional transcriptionist.

In the third phase, we approached three ongoing groups for discussion in focus groups. One of these was a specifically black group; one consisted of women studying at a residential college; one was a rural health group. This involved our recounting some of what we had found out—or rather, our interpretations of the material—and discussing it with the women concerned. This "member check" became for each of the groups an occasion for further reflection with one another and with us. We regard this as an important part of the research process. We also asked them to discuss some specially written "media accounts" featuring science (a composite made up from several sources). At a later date, a group of Muslim women was constituted to explore issues around Islam and science. We did so mainly because the way in which we selected people for interview (that is, on the basis of a sampling of groups and thereafter volunteers) actually screened out Asian women. Since they form a substantial part of the local population we wanted to rectify this skewing, at least to some degree.

All group discussions (which we conducted) were taped, listened to, and discussed by both of us soon afterwards. We shared the work of transcription, again inserting contextual notes.

We approached *groups*, different constituencies of women in different contexts, in an effort not to individualize women or de-contextualize the research too much. This involved fairly labor-intensive efforts to recruit volunteers for interviews. In some cases, this could have easily been achieved via tutors or group leaders, but we always attended class and groups meetings to explain what we were about. Black women's groups, in any case, made it clear that they would have insisted on our meeting them as a group to discuss the nature and possible usefulness of the research—even if we had not intended to do so! As soon as they were convinced that the research was worthwhile, they were happy to participate (see Cannon, Higginbotham, and Leung, 1991). Consistent with their stated views about science, the pursuit of knowledge (ours in this case) which would not serve some really useful purpose, was anathema to them. This did not mean it had to promise a direct pay-off for *them*, however.

That social relations of class, gender, and "race" are relevant to the production of knowledge—that they are "epistemological factors"—is a founding premise of our research. Our own experience as researchers and

as adult educators, as well as our reading of feminist theory and the social studies of science, convince us that this is so. It is obvious that our position as white, university-based female researchers had a bearing on the kind of conversations we had with the women, and so on the final product of the research.

In doing the research we have been influenced by the writings of Dorothy Smith (see 1979, 1987) and by her idea of feminist research as requiring the researcher to be located on the "same critical plane" as the women being researched. To the extent that this means not pretending to be able to achieve a "God's eye" view, this is what we did. We did not see our job as interpreting their testimony in terms of any fixed feminist, sociological, or other categories or theoretical projects. But our understandings and interpretations are clearly deeply influenced by our theoretical beliefs, our experiences as feminist women, and our being emotionally invested and contextually located (see Stanley and Wise, 1990, p. 39).

In both the research process and our writing up of the research, our aim is to move between different standpoints and contexts—between accounts of individual women, accounts of different groups, and what we say about them (influenced by our reading of feminist theory and our own experiences, and so on)—in an effort, not to arrive at some privileged account, but to produce a text, which, in doing justice to the women's own testimony, "exceeds our own understandings" (Lather, 1994, p. 7). Our strategy in analyzing the interviews, questionnaires, and group discussions is to seek a balance between identifying persistent themes across the interviews, charting differences between different groups, and treating each woman's narrative as a complete text.

We attempted to make the research as interactive and "power-sensitive" (Haraway, 1991a) as possible, notwithstanding the constraints of time and the "academic code of production" within which the research was located (see Stanley, 1990). In the end, of course, we have the pen: it is our job to give form to and interpret the data, and the women involved had the right to expect us to do our job.

Acker et al. (1983) have commented on the inappropriateness of just "being equal" in the relationship between researcher and researched, particularly when the researchers are university based. They suggest that attempts to create a more equal relationship can indeed become "exploitation and use," given that power differences cannot be eliminated. They found, too, that the women involved in their own research *wanted* sociological interpretations of their situation, seeing these as the outcome of special skills and knowledge to which they were not themselves privy.

This, then, is the paradox. But perhaps it arises at least partly out of assuming that the researcher holds all the cards. In fact, the women we interviewed in many cases carved out a space of their own, sometimes being quite playful—"If you *really* want a scientific answer, then. . . ."—or merely anticipating what they thought *we* expected. Thus, a frequent comment would be "I'm not sure if that's what you want" or "I hope I've been of some help." In these comments we are reminded of the self-reflexive nature of such research encounters and that power is not one-sided (see Lather, 1994). However, even if we acknowledge that "the interview always exceeds and transgresses our attempts to capture and categorize" (p. 28) this does not mean that we can avoid "the inevitable interpretive weight" which is on us as researchers (Lather, 1994, p. 8). That, after all, is the name of the game.

## The Women Interviewed

Brief biographical details of the women interviewed are given below. Details include age (in 1992), ethnicity, course studied or group attended, previous occupations, and occupations of parents (if this was mentioned in the interview). We have changed all names.

Alice, 52, white, educated in Australia; studying for Certificate in Literary and Cultural Studies, Warwick University; accountant; father an aviation store foreman, mother a machinist.

Anjana, 32, Ugandan Asian, came to England at age 13, in Access to Health course at Tile Hill College, Coventry; former pharmacy technician; father an accountant, mother a housewife.

Barbara, 32, brought up by her grandmother in Jamaica, came to Britain in 1970; member of Osaba Black Women's Group, studying school mathematics; Osaba playgroup leader; father a civil engineer, mother a nurse.

Bell, 26, black; in Access course at Hillcroft Residential College for Women; factory and travel agency work; father and mother factory workers.

Betsy, 45, black, from Grenada; member of Cariba Women's Group; nurse; father a shoemaker, mother a cleaner.

Brenda, 28, black, born and educated in Zimbabwe; in Access to Nursing course, Tile Hill College; various clerking and odd jobs; father a teacher, mother a nurse.

Carla, 46, white; in Women and Science course in Coventry; bar worker; father a French polisher, mother a housewife.

Cath, 42, white; in Access course at Hillcroft; shorthand typist and women's and peace movement voluntary worker; father a painter and decorator, mother a laundrette attendant.

Catherine, 49, white; in Women and Science course; housewife; father a toolmaker, mother a housewife.

Chris, 23, white; in Access course at Hillcroft; interior designer; father a sales and marketing manager, mother a personal assistant for a retail outlet.

Denise, 27, white; in Access course at Hillcroft as a day student; van driver and sales manager; father a Naval officer, mother a primary teacher.

Edna, 53, black, from Jamaica; member of Cariba Women's Group; nurse, then teacher; father an accountant, mother a housewife.

Elinor, 35, white; in Access course at Hillcroft as a day student; office work and catering; brought up by single mother with three brothers, who are now mechanics.

Gaynor, 32, white; member of Stoke-on-Trent Women and Health Group; former millworker, now runs a tearoom; father a millworker.

Gita, 29, Asian English, Sikh, parents came to England in the 1950s from north India; in Access to Health (Sports Science) course, Tile Hill College; shop work and care assistant; father a factory cleaner, mother looked after eleven children.

Gloria, 26, black, born severely disabled, now a wheelchair user; studying for Certificate in Literary and Cultural Studies, Warwick University (adult education); poet; father Zimbabwean, mother Polish, both University graduates.

Isabel, 38, white; in Access course at Hillcroft; army driver; father a joiner, mother a machinist and assistant cook.

Jackie, 30, white; in Access to Computing course, Tile Hill; semi-professional singer; father a computer engineer, mother a clerical assistant.

Janet, 48, white; member of Stoke-on-Trent Women and Health Group; shop assistant and millworker.

Jenny, 58, white; in Women and Science course; bookkeeper and tracer in car industry; father an accountant, mother a housewife.

Jo, 51, white; studying for Certificate in Literary and Cultural Studies, Warwick University; home economics teacher; father a factory storeman, mother a cook.

Kay, 42, white; member of Stoke-on-Trent Women and Health Group; lab technician and smallholder; father a canal maintenance worker, mother a housewife.

Kirsty, 28, white; in Access to Nursing course, Tile Hill; part-time bar work and shop work; father a journalist on local paper, mother housewife.

Loretta, 35, white; in Women and Science course; housewife.

Lynn, 66, white; member of Stoke-on-Trent Women and Health Group; retired teacher; father a minister, mother a housewife.

Martha, 57, black, from Guyana; member of Cariba Afro-Caribbean Women's Group, based in Coventry; nursing auxiliary; father a laborer, mother a schoolteacher.

Mary, 29; member of Osaba Black Women's Group; secretarial, then earned a degree, now Training and Development Officer for Osaba.

Meg, 48, white; in Women and Science course; voluntary school auxiliary; father in insurance, mother a housewife.

Michelle, 54, white; in Access course at Hillcroft; housekeeper for father, shop assistant.

Miranda, 37, white; in Community Access course, Tile Hill; nursing previously.

Molly, 35, black, brought up by grandmother in Barbados; in Access course at Hillcroft; single parent.

Monica, 36, white; studying for Certificate in Literary and Cultural Studies, Warwick University; clerk.

Rita, 28, Asian; in Access to Nursing course, Tile Hill; secretarial work then work with mentally handicapped adults and children.

Sally, 51, white; in Access course at Hillcroft; nanny and odd jobs; father in telecommunications, mother a housewife.

Sandra, 44, white; member of Stoke-on-Trent Women and Health Group; retired teacher; father a farmer, mother a housewife (and trained butcher).

Sandy, 28, white; in Women and Science course; shop and warehouse worker; father a shop convener, mother head of postroom.

Selma, 60, white Scottish, studying Certificate in Literary and Cultural Studies, Warwick University; various jobs, including shop assistant; father a carpenter and actor, mother a housewife and landlady (taking in lodgers).

Tania, 24, white; in Access course at Hillcroft; policewoman; father a postman, mother a payroll executive.

Terry, 44, white; studying for Certificate in Literary and Cultural Studies, Warwick University; dispenser; father a laborer, mother a dressmaker.

Veronica, 49, white; studying for Certificate in Literary and Cultural Studies, Warwick University; dressmaker, homeworker; father a brass caster, mother a housewife.

## Notes

1. These questions are modeled on ones used in an introductory science course for women, run by Shelagh Doonan. We are grateful to her for permission to use these questions here.

2. "Access" courses are intended to provide access to higher education for adults otherwise lacking in qualifications. They are usually based in colleges.

# Notes

## Introduction

1. The Mechanics' Institutes were established in the early nineteenth century as a means of providing education for working-class people (but primarily men). In some ways, these were radical, in that they sought to encourage debate and learning; yet—as many critics noted with fear—that debate could encourage dissent.

## Chapter One

2. See chapter in part 1 of Nancy Tuana, ed., 1989, *Feminism and Science* (Bloomington, Indiana University Press).
3. For example, see Ruth Hubbard, 1990, *The Politics of Women's Biology*; Anne Fausto-Sterling, 1992, *Myths of Gender*; Lynda Birke, 1986, *Women, Feminism, and Biology*; and Sue Rosser, 1992, *Biology and Feminism*. For historical analysis, see Londa Schiebinger, 1989, *The Mind Has No Sex?*.

## Chapter Four

4. It is, of course, possible that these nonverbal reactions resulted from the process of being interviewed. This may partly be the case, although the reactions were much less noticeable when women were talking about "safer" topics, such as their children or the course they were taking.
5. Bruno Latour makes a similar point, noting that the construction of the language and argument in a scientific paper "is chasing its readers away. . . . Made for attack and defence, it is no more a place for a leisurely stay than a bastion or a bunker. This makes it quite different from the reading of the Bible, Stendhal, or the poems of T. S. Eliot." Bruno Latour, *Science in Action*, 1987, (Milton Keynes, Open University Press), p. 61.
6. It has in fact been argued that the "nature of women's knowledge" makes it impossible to distinguish between what is emotional and what is

rational. Hilary Rose, for instance, has argued that women's "caring" work, even in the alienated forms generated by the social division of labor under capitalism, fosters a distinctive understanding of the social and material world which is more relational and governed by a "rationality of responsibility for others"—where "others" include the "natural" as well as the social world (Rose, 1994, p. 49; see also Hart, 1992). This "feminist reconstruction of rationality," which goes hand in hand with a "re-visioning" of the concept of labor so that emotion is restored within work and knowledge, is, she believes, a basis for a better and more accurate understanding of the social and natural world than the disembodied rationality of mainstream science.

This is a problematic argument since it relies so much on women's distinctive labor, particularly as mothers—an experience that is not shared by all. We also believe with Ruth Hubbard that it is the political understanding which feminism gives that leads to the possibility of a more accurate understanding of nature and that it is as political beings, not simply as women per se, that feminists and women will, if at all, change science.

## Chapter Five

7. Although we use the term *silence* here, no one was silent for very long. Our interpretation is based on what followed, usually a refusal to engage with particular ideas: for example, "physics? Oh, I don't know anything about that; I found it completely irrelevant and boring" might follow a short silence. Statements like that serve in conversation to deflect the interrogation and change the subject.

8. Even to the point where it was usually possible to recognize at a glance those particular parts of the interview transcripts, where otherwise articulate women suddenly lapsed into almost monosyllabic answers.

9. We recognize that silence as a strategy in these interviews must also be structured by our own positioning as white researchers who grew up in Britain.

10. This kind of claim is made, for instance, in relation to scientists' use of animals in research; "if only the public knew more" about why the research is done, then they would support it.

11. One example comes from the people living around the Deerfield river valley, in western Massachusetts, over the time of decommissioning the Yankee Rowe nuclear plant. The Citizens Awareness Network has been collecting health data (CAN, 1994), but how easy is it going to be to *prove* the radiation is causing all these effects?

12. Generally, lay people work with "an expanded vocabulary of risk" (Hornig, 1993); that is, they tend to reject the statistics as being insufficient and demand that risk assessments include the social aspects of risk. It is precisely these aspect that are missed in statistical/epidemiological data.

13. In arguing for more inclusive notions of rationality and science so as to include the experiences and understandings of excluded others, we are not of

course arguing that such experiences and understandings in themselves provide reliable *grounds* for knowledge claims about the world. Our "experience" often lies to us, not least because subordinated groups internalize what dominant groups believe about them. Nor would we want such arguments to be taken as justifying curricular and educational approaches which, in an effort to validate women's lived experiences, shortchange women by not moving outward from the confines of their own lives. Nevertheless, if knowledge is to be *created* from the perspective of many women's lives—"from below"—as "feminist standpoint" approaches propose (see ch. 7), the public act of women naming their experiences in their terms is a fundamental prerequisite. Otherwise, they will not see themselves as the kind of people who can make knowledge (see Harding, 1994, p. 20).

## Chapter Six

14. The heroism of the scientist is a theme in many films depicting specific scientists' lives in the cinema (Elena, 1993). No doubt this contributes to the impression of "gee-whiz" science.

15. "Oncomouse" was the name given by Harvard researchers to the strain of mouse bred for susceptibility to certain cancers. This strain became notorious because the researchers applied for patent rights. We use a journalistic version of the oncomouse story as a basis for part of the discussion in focus groups.

16. There is dispute over how it arose, but one suggestion is that it arose from giving cattle food containing offal from slaughterhouses, which included brains from sheep infected with the similar viral agent causing scrapie.

17. See Sandra Harding's 1993 collection *The "Racial" Economy of Science* (Bloomington, Indiana University Press) for further discussion of Eurocentrism in science.

## Chapter Seven

18. Also see Lynda Birke (1994) for an analysis of how this exclusion of caring and empathy form part of the training of biologists.

19. This is not to argue that women and men do think differently in some essentialist way. But, on the whole, highly gendered societies are likely to produce men and women who do have somewhat different views on the world.

20. Perhaps few *would* now dispute that scientific theories are forever "underdetermined" by all of the evidence we have or will ever have. Nelson (1990) also believes that many theories might indefinitely work equally well but that not all or any are compatible with our experience and other knowledge. So although we do construct theories and do not discover them, and although they are forever underdetermined by the evidence "it is not up to us whether gravity is real or whether many research programmes are andro-

centric" (p. 295). Experience indicates that some theories *are* better than others (p. 34)—as attempts to suspend belief in gravity show!

21. A central weakness of postmodernist-inspired feminism, suggests a recent critique, is its assumption that either we must accept the abstract, context-independent, disembodied individual of traditional epistemology (and of the "common sense" of science) as the paradigm "knower" *or* concede that as "subjects in process," constituted by "the discourses and practices of their culture," persons cannot be subjects or agents of knowledge in the sense (it is argued) epistemology requires (see Tuana, 1992; Nelson, 1993).

22. Nor should we underestimate the role of emotions in assisting or hindering feminist projects. The recent flowering of feminist epistemology around critiques of science could itself be hindered in its efforts to contribute to its emancipatory goal if it fell into the trap of "rationalism" (see Tuana, 1992).

23. As Hilary Rose observes (1994, p. 102 ff.), women are in fact overrepresented in some parts of science—as they are in other parts of the highly segregated labor market that characterizes our domestic economy and the "new international division of labor." They are present in huge numbers as the "underlaborers" of science (paralleling their "primary" tasks in the home) but are absent, as elsewhere, from science's "centers of power" (Cacoullis, quoted in Rose, 1994, p. 102). Indeed, Rose argues that it is likely that the relationship between two systems of production—the production of things or commodities (for profit) and the production of people (for life)—holds the key to understanding why there are so few women in science *and* why the knowledge produced by science is so abstract and disembodied (see Rose, p. 22 ff.; see also, Hart, 1992, esp. pp. 118 ff. for a similar argument and an examination of its implications for adult education)

## Chapter Eight.

24. The science/not science boundary needs to be challenged rather than taken for granted. This might be done, for example, by the inclusion by feminists working within scientific disciplines of literary analyses of scientific texts (something that is fairly common outside scientific disciplines but rare among those who have a detailed knowledge of science from the inside).

It is equally important for feminists working in the humanities to attempt to engage with the sciences rather than reject "science" outright—a "them and us" approach that leaves the authority enjoyed by the sciences essentially unchallenged and, indeed, misunderstood. This "them and us" approach not only fails to understand how the power and authority of different sciences is constructed (differently); it also fails to acknowledge points of resistance within the sciences themselves to "science's" own systems of authority.

Acker, Joan, Kate Barry, and Joke Esseveld. 1983. "Objectivity and Truth: Problems in Doing Feminist Research." *Women's Studies International Forum* 6, 423–35.

Addelson, K. P. 1983. "The Man of Professional Wisdom." In S. Harding and M. Hintikka, eds. (1983).

Afzal-ur-Rahman. 1981. *Quranic Sciences*. London: Muslim Schools Trust.

Ahmed, Leila. 1992. *Women and Gender in Islam: Historical Roots of a Modern Debate*. New Haven: Yale University Press.

al-Hassan, Ahmed Y. See Hassan

Alexander, D. 1994. "The Education of Adults in Scotland: Democracy and Curriculum." *Studies in the Education of Adults* 26, 31–49.

Arendt, H. 1968. Introduction. In W. Benjamin, *Illuminations*, trans. Harry Zohn. New York: Harcourt, Brace, and World.

Barad, Karen. 1995. "A Feminist Approach to Teaching Quantum Physics." In Rosser, ed. (1995).

Barr, Jean. 1984. "Women's Education—The Ways Forward." In *The Future of Women's Education*. Conference Report, Durham University.

———. 1991. "Women, Education, and Counselling." *Reportback* 1 (Workers Educational Assoc.), 20–21.

———, and Lynda Birke. 1994. "Women, Science, and Adult Education: Towards a Feminist Curriculum?" *Women's Studies International Forum* 17, 473–83.

Barr, Mohamed Ali. 1986. *Human Development as Revealed in the Holy Quran and Hadith*. Jeddah: Saudi Publishing and Distribution House.

Bauer, Martin, and I. Schoon. 1993. "Mapping Variety in Public Understanding of Science." *Public Understanding of Science* 2, 141–55.

Beck, Ulrich. 1992. *Risk Society: Towards a New Modernity*, trans. Mark Ritter. London: Sage.

Belanger, Paul. 1994. "Lifelong Learning: The Dialectics of 'Lifelong Education.'" *International Review of Education* 40 (3–5), 353–81.

Belenky, M. F., B. M. Clinchy, N. R. Goldberger, and J. M. Tarule. 1986. *Women's Ways of Knowing: The Development of Self, Voice, and Mind*. New York: Ba-sic.

Bell, Susan. 1994. "Science for the People?" *Women's Studies International Forum* 17.

Bhaskar, R. 1989. *Reclaiming Reality: A Critical Approach to Contemporary Philosophy*. London: Verso.

Biology and Gender Study Group. 1989. "The Importance of Feminist Critique for Contemporary Cell Biology." In N. Tuana, ed., *Feminism and Science*. Bloomington: Indiana University Press.

Birke, Lynda. 1986. *Women, Feminism, and Biology: The Feminist Challenge*. Brighton: Wheatsheaf.

———. 1991. "Adult Education and the Public Understanding of Science." *Journal of Further and Higher Education* 15, 15–23.

———. 1992. "Inside Science for Women: Common Sense or Science?" *Journal of Further and Higher Education* 16, 18–29.

———. 1994. *Feminism, Animals, and Science: The Naming of the Shrew*. Buckingham: Open University Press.

———, and C. Dunlop. 1993. "Bringing Science to Women?" *Adults Learning* (March).

Block, Jonathan. 1994. "Another Fine Year at Vermont Yankee: $237,500.00." *On the Watch*. New England Coalition on Nuclear Pollution Newsletter (Spring).

Blundell, S. 1992. "Gender and the Curriculum in Adult Education." *International Journal of Lifelong Education* 11, 199–216.

Bodmer, Walter, and Robin McKie. 1994. *The Book of Man: The Quest to Discover our Genetic Heritage*. London: Little, Brown.

Bordo, S. 1990. "Feminism, Postmodernism, and Gender-Scepticism." In L. Nicholson, ed., *Feminism/Postmodernism*. London: Routledge, 133–56.

Boxer, Marilyn. 1988. "For and about Women: The Theory and Practice of Women's Studies in the United States." In E. Minnich, J. O'Barr, and R. Rosenfeld, eds., *Reconstructing the Academy: Women's Education and Women's Studies*. Chicago: University of Chicago Press.

Braidotti, Rosi. 1994. *Nomadic Subjects: Embodiment and Sexual Difference in Contemporary Feminist Theory*. New York: Columbia University Press.

———, E. Charkiewicz, S. Hausler, and S. Wieringa. 1994. *Women, the Environment, and Sustainable Development: Towards a Theoretical Synthesis*. London: Zed.

Brighton Women and Science Group. 1980. *Alice through the Microscope: The Power of Science over Women's Lives*. London: Virago.

Brookfield, S. 1989. "The Epistemology of Adult Education in the U.S. and G.B.: A Cross-Cultural Analysis." In Barry Bright, ed., *Theory and Practice in the Study of Adult Education: The Epistemological Debate*. London: Routledge, 143–73.

Butler, Sandra, and Claire Wintram. 1991. *Feminist Groupwork*. London: Sage.

Cameron, Lynn, Elizabeth Higginbottom, and Marianne Leung. 1988. "Race

and Class Bias in Qualitative Research on Women." *Gender and Society* 2, 449–62.
Campbell, M. A., and R. K. Campbell-Wright. 1995. "Toward a Feminist Algebra." In S. Rosser, ed. (1995).
Chalmers, A. F. 1982. "Epidemiology and the Scientific Method." *International Journal of Health Studies* 12 (4).
Charles, N. and M. Kerr. 1988. *Women, Food and Families.* Manchester: Manchester University Press.
Chow, Rey. 1989. "'It's You and Not Me': Domination and 'Othering' in Theorizing the 'Third World.'" In E. Weed, ed., *Coming to Terms: Feminism, Theory, Politics*. London: Routledge.
Citizens Awareness Newsletter (CAN). 1994. "The Carcinogenic, Mutagenic, Teratogenic, and Transmutational Effects of Tritium" (April).
Cockburn, Cynthia. 1985. *Machinery of Dominance: Women, Men, and Technical Know-How*. London: Pluto.
Code, Lorraine. 1989. "Experience, Knowledge, and Responsibility." In Ann Garry and Marilyn Pearsall, eds., *Women, Knowledge, and Reality*. London: Unwin Hyman, 157–72.
Collins, Harry M. 1985. *Changing Order: Replication and Induction in Scientific Practice*. London: Sage.
Collins, Patricia Hill. 1990. *Black Feminist Thought: Knowledge, Consciousness, and the Politics of Empowerment*. London: Unwin Hyman.
Cowburn, Will. 1986. *Class, Ideology, and Community Education*. London: Croom Helm.
Elena, Alberto. 1993. "Exemplary Lives: Biographies of Scientists on the Screen." *Public Understanding of Science* 2, 205–23.
Fairclough, J. 1989. *Language and Power.* London: Longman.
Fausto-Sterling, Anne. 1992. *Myths of Gender: Biological Theories about Women and Men,* 2d ed. New York: Basic.
Fay, Brian. 1987. *Critical Social Science: Liberation and Its Limits*. Cambridge: Polity.
Felman, Shoshana. 1982. "Psychoanalysis and Education: Teaching Terminable and Interminable." In B. Johnson, ed., *The Pedagogical Imperative: Teaching as a Literary Genre*. New Haven: Yale University Press, 21–44.
———, and D. Laub. 1992. *Testimony: Crises of Witnessing in Literature, Psychoanalysis, and History*. London: Routledge.
Fiddes, Nick. 1991. *Meat: A Natural Symbol*. London: Routledge.
Fine, Michelle, and Nancie Zane. 1991. "Bein' Wrapped Up Too Tight: When Low-Income Women Drop Out of High School." *Women's Studies Quarterly* 19, 77–99.
Foucault, Michel. 1979. *Discipline and Punish: The Birth of the Prison,* trans. Alan Sheridan. New York: Vintage.
Freire, Paulo. 1970. *Pedagogy of the Oppressed,* trans. Myra B. Ramos. New York: Continuum.
———. 1983. "Education and Conscientizaçâo." In Malcolm Tight, ed., *Adult Learning and Education*. London: Croom Helm.

French, J. 1989. "Accomplishing Scientific Instruction." In R. Millar, ed., *Doing Science: Images of Science in Science Education*. London: Falmer.
Gatens, Moira. 1991. *Feminism and Philosophy: Perspectives on Difference and Equality*. Cambridge: Polity.
———. 1992. "Power, Bodies, and Difference." In Michele Barrett and Anne Phillips, eds., *Destabilizing Theory: Contemporary Feminist Debates*. Cambridge: Polity, 120–37.
Gibbons, Michael, C. Limoges, H. Nowotny, S. Schwartzman, P. Scott, and M. Trow. 1994. *The New Production of Knowledge: The Dynamics of Science and Research in Contemporary Societies*. London: Sage.
Giddens, Anthony. 1990. *The Consequences of Modernity*. Cambridge: Polity.
Gilligan, Carol. 1982. *In a Different Voice: Psychological Theory and Women's Development*. Cambridge, Mass.: Harvard University Press.
Ginzburg, Carlo. 1980. "Morelli, Freud, and Sherlock Holmes: Clues and Scientific Method." *History Workshop Journal*, 5–36.
Goldthorpe, John H., Catriona Llewellyn, and Clive Payne. 1987. *Social Mobility and Class Structure in Modern Britain,* 2nd ed. Oxford: Clarendon Press.
Goodman, David, and Michael Redclift. 1991. *Refashioning Nature: Food, Ecology, and Culture*. London: Routledge.
Grant, Linda. 1994. "First among Equals." *Guardian Weekend*, 22/10/94.
Gross, M., and M. B. Averill. 1983. "Evolution and Patriarchal Myths of Scarcity and Competition." In S. Harding and M. Hintikka, eds. (1983).
Grossman, Karl. 1993. "Environmental Racism." In S. Harding, ed. (1993).
Haraway, Donna J. 1989. *Primate Visions: Gender, Race, and Nature in the World of Modern Science*. London: Routledge.
———. 1991a. "Situated Knowledges: The Science Question in Feminism and the Privilege of Partial Perspective." In D. Haraway, *Simians, Cyborgs and Women: The Reinvention of Nature.* London: Routledge, 183–201.
———. 1991b. "The Contest for Primate Nature: Daughters of Man-the-Hunter in the Field, 1960–80." In D. Haraway, *Simians, Cyborgs, and Women: The Reinvention of Nature*. London: Routledge, 81–108.
Harding, Sandra. 1986. *The Science Question in Feminism*. Milton Keynes: Open University Press.
———. 1991. *Whose Science? Whose Knowledge? Thinking from Women's Lives*. Milton Keynes: Open University Press.
———. 1992. "How the Women's Movement Benefits Science." In G. Kirkup and L. Smith Keller, eds. (1992).
———. 1994a. "Is Science Multicultural? Challenges, Resources, Opportunities, Uncertainties." *Configurations* 2, 301–30.
———. 1994b. "Subjectivity, Experience, and Knowledge: An Epistemology from/for Rainbow Coalition Politics." In Judith Roof and Robyn Wiegman, eds., *Who Can Speak: Authority and Critical Identity*. Urbana: University of Illinois Press.
———, ed. 1993. *The "Racial" Economy of Science: Toward Democratic Future*. Bloomington: Indiana University Press.
———, and Merrill B. Hintikka, eds. 1983. *Discovering Reality: Feminist Per-*

*spectives on Epistemology, Metaphysics, Methodology, and Philosophy of Science.* Dordrecht: Reidel.

Harré, Rom. 1986. *Varieties of Realism: A Rationale for the Natural Sciences.* London: Blackwell.

Hart, Mechtild. 1992. *Working and Educating for Life: Feminist and International Perspectives on Adult Education.* London: Routledge.

al-Hassan, Ahmed Y., and Donald R. Hill. 1986. *Islamic Technology: An Illustrated History.* Cambridge: Cambridge University Press.

Haynes, Roslynn D. 1994. *From Faust to Strangelove: Representations of the Scientist in Western Literature.* Baltimore: Johns Hopkins University Press.

hooks, bell. 1990. *Yearning: Race, Gender and Cultural Politics.* London: Turnaround.

———. 1994. *Teaching to Transgress: Education as the Practice of Freedom.* New York: Routledge.

Hornig, Susanna. 1993. "Reading Risk: Response to Print Media Accounts of Technological Risk." *Public Understanding of Science* 2, 95–109.

Hubbard, Ruth. 1990. *The Politics of Women's Biology.* New Brunswick: Rutgers University Press.

———, and Elijah Wald. 1993. *Exploding the Gene Myth: How Genetic Information is Produced and Manipulated by Scientists, Physicians, Employers, Insurance Companies, Educators, and Law Enforcers.* Boston: Beacon.

Jaggar, A. M. 1989. "Love and Knowledge: Emotion in Feminist Epistemology." In A. Garry and M. Pearsall, eds., *Women, Knowledge and Reality: Explorations in Feminist Philosophy.* London: Unwin Hyman, 129–55.

James, Allison. 1993. "Eating Greens: Discourses of Organic Food." In K. Milton, ed., *Environmentalism: The View from Anthropology.* London: Routledge.

Johnson, R. 1988. "Really Useful Knowledge, 1790–1850." In T. Lovett, ed., *Radical Approaches to Adult Education: A Reader.* London: Routledge, 3–34.

Keddie, Nell. 1981. "Adult Education: A Women's Service." Unpublished paper.

Keller, Evelyn Fox. 1983. *A Feeling for the Organism: The Life and Work of Barbara McClintock.* San Francisco: Freeman.

———. 1985. *Reflections on Gender and Science.* New Haven: Yale University Press.

———. 1992. "How Gender Matters or Why It Is So Hard for Us to Count Past Two." In G. Kirkup and L. Smith Keller, eds. (1992), 42–56.

Keller, L. A. 1992. "Discovering and Doing: Science and Technology, an Introduction." In G. Kirkup and L. Smith Keller, eds.

Kirkup, G. and L. Smith Keller, eds. 1992. *Inventing Women.* Cambridge: Polity.

Kitzinger, Celia, and Rachel Perkins. 1993. *Changing Our Minds: Feminist Transformations of Knowledge.* London: Onlywomen.

Knorr-Cetina, Karin D. 1983. "The Ethnographic Study of Scientific Work: Towards a Constructionist Interpretation of Science." In K. Knorr-Cetina and M. Mulkay, eds., *Science Observed: Perspectives on the Social Study of Science.* London: Sage.

Knowles, Malcolm S. 1978. *The Adult Learner: A Neglected Species,* 2d ed. Houston: Gulf.

Kuhn, Annette. 1994. *Women's Pictures: Feminism and Cinema,* 2d ed. London: Verso.

Langer, Susanne K. 1988. *Mind: An Essay on Human Feeling*, abridged ed. Baltimore: Johns Hopkins University Press.

Larochelle, Marie, and Jacques Desantels. 1991. "Of Course, It's Just Obvious: Adolescents' Ideas of Scientific Knowledge." *International Journal of Science Education* 13, 373–89.

Lather, Patti. 1991. *Getting Smart: Feminist Research and Pedagogy with/in the Postmodern*. London: Routledge.

———. 1994. "Textuality as Praxis." Paper presented to a meeting of the American Educational Research Association, New Orleans.

Latour, Bruno. 1987. *Science in Action: How to Follow Scientists and Engineers through Society*. Milton Keynes: Open University Press.

Layton, David. 1973. *Science for the People*. London: Allen and Unwin.

———, A. Davey, and E. Jenkins. 1986. "Science for Specific Social Purposes (SSSP): Perspectives on Adult Scientific Literacy." *Studies in Science Education* 13, 27–52.

———, E. Jenkins, S. Macgill, and A. Davey. 1993. *Inarticulate Science? Perspectives on the Public Understanding of Science and Some Implications for Science Education*. Driffield: Studies in Education.

Le Doeuff, Michele. 1991. *Hipparchia's Choice: An Essay Concerning Women, Philosophy, etc*. Oxford: Blackwell.

Lewis, Magda Gere. 1993. *Without a Word: Teaching beyond Women's Silence*. London: Routledge.

Lloyd, Genevieve. 1984. *The Man of Reason: "Male" and "Female" in Western Philosophy*. London: Methuen.

Longino, Helen E. 1990. *Science as Social Knowledge: Values and Objectivity in Scientific Inquiry*. Princeton: Princeton University Press.

Lugones, Maria C., and Elizabeth V. Spelman. 1983. "Have We Got a Theory For You? Feminist Theory, Cultural Imperialism and the Demand for the Woman's Voice." *Women's Studies International Forum* 6, 573–81.

Luttrell, Wendy. 1989. "Working Class Women's Ways of Knowing: Effects of Gender, Race, and Class." *Sociology of Education* 62, 33–46.

Lyotard, Jean Francois. 1984. *The Postmodern Condition: A Report on Knowledge*, trans. Geoff Bennington and Brian Massumi. Minneapolis: University of Minnesota Press.

Martin, Emily. 1989. *The Woman in the Body*. Boston: Beacon.

———. 1994. *Flexible Bodies: Tracking Immunity in American Culture from the Days of Polio to the Age of AIDS*. Boston: Beacon.

McNeil, Maureen, ed. 1987. *Gender and Expertise*. London: Free Association.

Merchant, Caroline. 1980. *The Death of Nature: Women, Ecology, and the Scientific Revolution*. San Francisco: Harper and Row.

Messing, Karen, and Donna Mergler. 1995. "The Rat Couldn't Speak but We Can: Inhumanity in Occupational Health Research." In L. Birke and R. Hubbard, eds., *Reinventing Biology: Respect for Life and the Creation of Knowledge*. Bloomington: Indiana University Press.

Michael, M. 1992. "Lay Discourses of Science: Science-in-General, Science-in-Particular, and Self." *Science, Technology, and Human Values* 17, 313–33.

———. 1995. "Knowing Ignorance and Ignoring Knowledge: Discourse of Ignorance in the Public Understanding of Science and Technology." In A. Irwin and B. Wynne, eds., *Science, Technology, and Everyday Life* (forthcoming).

Midgley, Mary. 1992. *Science as Salvation: A Modern Myth and Its Meaning.* London: Routledge.

Millar, Robin, and R. Driver. 1987. "Beyond Process." *Studies in Science Education* 14, 33–62.

Miller, J. D. 1993. "Theory and Measurement in the Public Understanding of Science: A Rejoinder to Bauer and Schoon." *Public Understanding of Science* 2, 235–43.

Minnich, Elizabeth. 1990. *Transforming Knowledge.* Philadelphia: Temple University Press.

Moheno, P. B. B. 1993. "Toward a Fully Human Science Education: An Exploratory Study of Prospective Teachers' Attitudes toward Humanistic Science Education." *International Journal of Science Education* 15, 95–106.

Moi, Toril. 1989. "Patriarchal Thought and the Drive for Knowledge." In T. Brennan, ed., *Between Feminism and Psychoanalysis.* London: Routledge.

Montgomery, Scott L. 1994. *Minds for the Making: The Role of Science in American Education, 1750–1990.* New York: Guilford.

Morrison, Toni. 1994. *Playing in the Dark: Whiteness and the Literary Imagination.* Cambridge, Mass.: Harvard University Press.

Nandy, Ashis. 1988. "Science as a Reason of State." In A. Nandy, ed., *Science, Hegemony, and Violence: A Requiem for Modernity.* New Delhi: Oxford University Press.

Needham, Joseph. 1993. "Poverties and Triumphs of the Chinese Scientific Tradition." In S. Harding, ed. (1993), 30–46.

Nelkin, Dorothy. 1987. *Selling Science: How the Press Covers Science and Technology.* New York: Freeman.

Nelson, Lynn. H. 1990. *Who Knows: From Quine to Feminist Empiricism.* Philadelphia: Temple University Press.

———. H. 1993. "Epistemological Communities." In L. Alcoff and E. Potter, eds., *Feminist Epistemologies.* London: Routledge, 121–59.

Pagano, Jo Anne. 1991. "Moral Fictions: The Dilemma of Theory and Practice." In C. Witherell and N. Noddings, eds., *Stories Lives Tell.* New York: Teachers College Press.

Parrish, Geov. 1994. "Burial Grounds: Toxic Waste in Native America." *On Indian Land* (Spring), 11.

Rich, Adrienne C. 1979. *On Lies, Secrets, and Silence: Selected Prose*, 1966–78. New York: Norton.

Riley, Denise. 1988. *Am I That Name? Feminism and the Category of "Women" in History.* London: Macmillan.

Rose, Hilary. 1994. *Love, Power, and Knowledge: Towards a Feminist Transformation of the Sciences.* Cambridge: Polity.

Rosser, Sue V. 1992. *Biology and Feminism: A Dynamic Interaction.* New York: Twayne.

———, and Bonnie Kelly. 1994. *Educating Women for Success in Science and Mathematics.* Columbia, S.C.: University of South Carolina.

———, ed. 1995. *Teaching the Majority: Breaking the Gender Barrier in Science, Mathematics, and Engineering.* New York: Teachers College Press.

Rossiter, Margaret W. 1982. *Women Scientists in America: Struggles and Strategies to 1940.* Baltimore: Johns Hopkins University Press.

Russell, T., and H. Munby. 1989. "Science as a Discipline, Science as Seen by Students and Teachers' Professional Knowledge." In R. Millar, ed., *Doing Science: Images of Science in Science Education.* London: Falmer.

Sacks, K. B. 1993. "Euro-Ethnic Working Class Women's Community Culture." *Frontiers* 14, 1–23.

Said, Edward. 1986. "Orientalism Reconsidered." In F. Barter et al., eds., *Liberation, Politics, and Theory.* New York: Methuen.

———. 1993. *Culture and Imperialism.* New York: Knopf.

Sayers, Janet. 1983. "Feminism and Science: Reason and Passion." *Women's Studies International Forum* 16, no. 4, 423–35.

Schama, Simon. 1995. *Landscape and Memory.* New York: Knopf.

Schiebinger, Londa L. 1989. *The Mind Has No Sex? Women in the Origins of Modern Science.* Cambridge, Mass.: Harvard University Press.

———. 1993. *Nature's Body: Sexual Politics and the Making of Modern Science.* Boston: Beacon.

Shepherd, Linda J. 1993. *Lifting the Veil: The Feminine Face of Science.* London: Shambhala.

Shiva, Vandana. 1988. *Staying Alive: Women, Ecology, and Development.* London: Zed.

Smith, Dorothy E. 1979. "A Sociology for Women." In J. A. Sherman and E. T. Beck, eds., *The Prism of Sex: Essays in the Sociology of Knowledge.* Madison: University of Wisconsin Press.

———. 1987. *The Everyday World as Problematic: A Feminist Sociology.* Milton Keynes: Open University Press.

Smithson, Michael. 1989. *Ignorance and Uncertainty: Emerging Paradigms.* New York: Springer-Verlag.

Solomon, Joan. 1992. "The Asian Girl and Her Science Education." Oxford: University of Oxford Department of Educational Studies, Report to British Telecom. December.

———. 1993. "Reception and Rejection of Science Knowledge: Choice, Style, and Home Culture." *Public Understanding of Science* 2, 111–21.

Spelman, Elizabeth. 1988. *Inessential Woman: Problems of Exclusion in Feminist Thought.* Boston: Beacon.

Spivak, Gayatry. 1987. *In Other Worlds: Essays in Cultural Politics.* London: Methuen.

Stanley, Liz. 1990. "Feminist Praxis and the Academic Mode of Production." In L. Stanley, ed. (1990).

———, and S. Wise. 1990. "Method, Methodology, and Epistemology in Feminist Research Processes." In L. Stanley, ed. (1990), 20–60.

———, ed., *Feminist Praxis: Research, Theory, and Epistemology in Feminist Sociology*. London: Routledge.
Stepan, Nancy L., and Sander L. Gilman. 1993. "Appropriating the Idioms of Science: The Rejection of Scientific Racism." In Sandra Harding, ed. (1993).
Stewart, J. 1988. "Science Shops in France: A Personal View." *Science as Culture* 2, 52–74.
Stocking, Holly S., and Lisa W. Holstein. 1993. "Constructing and Reconstructing Scientific Ignorance." *Knowledge: Creation, Diffusion, Utilization* 15, 186–210.
Third World Network. 1993. "Modern Science in Crisis: A Third World Response." In S. Harding, ed. (1993).
Thomas, Kim. 1990. *Gender and Subject in Higher Education*. Milton Keynes: Open University Press.
Thomas, Keith. 1983. *Man and the Natural World: Changing Attitudes in England, 1500–1800*. London: Allen Lane.
Thompson, Jane L. 1983. *Learning Liberation: Women's Responses to Men's Education*. London: Croom Helm.
———. 1993. "Learning Liberation: An Open Letter to Whoever's Left." *Adults Learning* 4, 44.
Tobach, Ethel, and Betty Rosoff, eds. 1994. *Challenging Racism and Sexism: Alternatives to Genetic Explanations*. New York: Feminist.
Tuana, Nancy 1989. "The Weaker Seed: The Sexist Bias of Reproductive Theory." In N. Tuana, ed., *Feminism and Science*. Bloomington: Indiana University Press.
———. 1992. "The Radical Future of Feminist Empiricism." *Hypatia* (Winter), 100–14.
Walkerdine, Valerie, and the Girls and Mathematics Unit. 1989. *Counting Girls Out*. London: Virago.
Wallis, R. 1985. "Science and Pseudoscience." *Information sur les Sciences Sociales* 24, 585–601.
Wynne, B. E. 1991. "Knowledges in Context." *Science, Technology and Human Values* 16, 111–21.
———. 1992. "Public Understanding of Science Research: New Horizons or Hall of Mirrors?" *Public Understanding of Science* 1, 37–43.

Common Science?

# Name Index

**JEAN BARR** COORDINATES POSTGRADUATE COURSES IN ADULT AND CONTINUING EDUCATION AT GLASGOW UNIVERSITY, WHERE SHE ALSO TEACHES COURSES IN THE PHILOSOPHY OF SOCIAL SCIENCE AND IN FEMINIST EDUCATION. SHE HAS WRITTEN ON EDUCATIONAL POLICY AND PRACTICE FOR SEVERAL INTERNATIONAL PUBLICATIONS.

**LYNDA BIRKE** IS SENIOR LECTURER AT THE CENTRE FOR THE STUDY OF WOMEN AND GENDER AT THE UNIVERSITY OF WARWICK AND A BIOLOGIST WHO HAS WRITTEN EXTENSIVELY ON FEMINISM AND SCIENCE. SHE IS THE COEDITOR OF *REINVENTING BIOLOGY* AND AUTHOR OF *FEMINISM, ANIMALS, AND SCIENCE; TOMORROW'S CHILD;* AND *WOMEN, FEMINISM, AND BIOLOGY.*